Generalized Network Improvement and Packing Problems

Michael Holzhauser

Generalized Network Improvement and Packing Problems

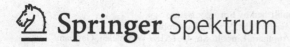

Michael Holzhauser
Technische Universität Kaiserslautern
Germany

Vom Fachbereich Mathematik der Technischen Universität Kaiserslautern zur Verleihung des akademischen Grades Doktor der Naturwissenschaften (Doctor rerum naturalium, Dr. rer. nat.) genehmigte Dissertation, 2016

D 386

Erstgutachter: Prof. Dr. Sven O. Krumke
Zweitgutachter: Prof. Dr. Andreas Bley
Tag der Disputation: 19. August 2016

ISBN 978-3-658-16811-7   ISBN 978-3-658-16812-4   (eBook)
DOI 10.1007/978-3-658-16812-4

Library of Congress Control Number: 2016961699

Springer Spektrum
© Springer Fachmedien Wiesbaden GmbH 2016

This Springer Spektrum imprint is published by Springer Nature
The registered company is Springer Fachmedien Wiesbaden GmbH
The registered company address is: Abraham-Lincoln-Str. 46, 65189 Wiesbaden, Germany

Acknowledgments

First and foremost, I am most indebted to my supervisor Prof. Dr. Sven O. Krumke, who gave me the opportunity to work in the Optimization Research Group at the University of Kaiserslautern and to write this thesis. Much more than this, I thank him for the never-ending support he gave me and the pleasant atmosphere when working together. His unlimited treasure of ideas and his overview of related work contributed significantly to the results of this thesis.

Furthermore, I would like to thank my co-supervisor Juniorprof. Dr. Clemens Thielen for all the support he gave me during the last three years and the dozens of hours we spent together while publishing our results. I also thank Prof. Dr. Andreas Bley for his effort in serving as a co-referee for this thesis.

I wish to thank all my colleagues in the Optimization Research Group for all the leisure time spent during lunch and coffee breaks and the interesting discussions. In particular, I want to express my gratitude to André Chassein, Michael Hopf, Lena Leiß, and Marco Natale (in alphabetical order) for their great support in proofreading parts of this thesis. Moreover, I thank the members of our Pub-Quiz team for all the evenings spiked with triumphs and full of dismal failures.

Last but not least, I would like to thank my parents, grandparents, and my brother for their never-ending support as well as Bernadette for all the patience and for making my life more beautiful.

Michael Holzhauser

Abstract

Network flow problems and packing problems in general are two of the most investigated classes of problems in discrete optimization. In the last decades, many combinatorial algorithms have been developed and steadily improved that compute exact or approximate solutions for these problems. Although these algorithms allow to solve many real world applications, most of them are highly tailored to the inherent structure of the underlying problems and do not admit even slight variations or extensions to these.

In this thesis, we investigate such extensions and variations of known network flow and packing problems with respect to their complexity and approximability. In the *budget-constrained minimum cost flow problem*, one seeks to determine a minimum cost flow subject to a budget constraint based on a second kind of costs. For this problem, we study efficient exact and approximate combinatorial algorithms. We also investigate two discrete variants, which can be interpreted as *network improvement problems* in which the edge capacities in the underlying network are allowed to be modified. Although the problem becomes hard to solve in these discrete settings, we are able to derive exact and approximate algorithms by exploiting an interesting connection to a novel variant of the traditional knapsack problem.

We also investigate two extensions of the traditional maximum flow problem. In the *maximum flow problem in generalized processing networks*, the aim is to determine a maximum flow in which the flow on each edge is additionally bounded by a *dynamic capacity* that depends on the total amount of flow leaving the starting node of the edge. Although this problem is as hard to solve as any linear fractional packing problem, we are able to adapt algorithms for the traditional maximum and minimum cost flow problem. Finally, we investigate an extension of the traditional maximum flow problem in which the flow leaving an edge is described by a convex increasing function of the flow entering the edge. While the problem becomes hard to solve and approximate even in its most simple form, we are able to derive exact algorithms.

Beyond these problems, we investigate the connection between network flow problems and packing problems in general by extending a well-known framework for deriving efficient approximation algorithms for packing problems to a large class of network flow problems.

Zusammenfassung

Netzwerkfluss- und Packungsprobleme gehören zu den meistuntersuchten Problemklassen der diskreten Optimierung. In den vergangenen Jahrzehnten wurde eine Vielzahl an kombinatorischen Algorithmen zur Berechnung von exakten oder approximativen Lösungen entwickelt und stetig weiterentwickelt. Obwohl sich diese Algorithmen auf eine große Zahl von realen Problemen anwenden lassen, sind die meisten stark auf die Struktur des zugrunde liegenden Problems zugeschnitten und erlauben keine Abänderungen oder Erweiterungen desselben.

In dieser Arbeit werden solche Erweiterungen und Variationen von bekannten Netzwerkfluss- und Packungsproblemen hinsichtlich ihrer Komplexität und Approximierbarkeit untersucht. Ziel im *budgetrestringierten Minimalkostenflussproblem* ist es, einen Minimalkostenfluss zu bestimmen, welcher zusätzlich durch eine Budgetbedingung basierend auf einem zweiten Kostentyp beschränkt ist. Für dieses Problem werden effiziente exakte und approximative kombinatorische Algorithmen präsentiert. Zudem werden zwei diskrete Varianten desselben Problems untersucht, welche als *Netzwerkausbauprobleme* interpretiert werden können, in denen die Kantenkapazitäten des zugrunde liegenden Netzwerks modifiziert werden können. Obwohl das Problem in diesen Varianten vom Standpunkt der Komplexitätstheorie schwer zu lösen ist, leiten wir exakte und approximative Lösungsverfahren her, indem unter anderem eine interessante Verbindung zu einer neuartigen Variante des Rucksackproblems aufgezeigt wird.

Des Weiteren werden zwei Erweiterungen des traditionellen Maximalflussproblems untersucht. Im *Maximalflussproblem in verallgemeinerten Prozessnetzwerken* ist das Ziel einen maximalen Fluss zu bestimmen, bei welchem der Fluss auf den Kanten zusätzlich durch eine *dynamische Kapazität* beschränkt ist, abhängig vom Gesamtfluss, der den Startknoten der Kante verlässt. Obwohl dieses Problem so schwer zu lösen ist wie jedes lineare fraktionale Packungsproblem, ist es möglich, Algorithmen für das Maximalfluss- und Minimalkostenflussproblem zu adaptieren. Schließlich werden wir eine Erweiterung des traditionellen Maximalflussproblems untersuchen, in dem der Fluss, der eine Kante verlässt, durch eine konvexe steigende Funktion in Abhängigkeit des einkommenden Flusses beschrieben ist. Für dieses Problem, welches schwer zu lösen und zu approximieren ist, werden exakte Algorithmen präsentiert.

Neben den genannten Problemen wird der Zusammenhang zwischen Netzwerkflussproblemen und Packungsproblemen im Allgemeinen untersucht, indem ein weit bekanntes Framework zum Ableiten von effizienten Approximationsalgorithmen für Packungsprobleme auf eine große Klasse von Netzwerkflussproblemen erweitert wird.

Contents

List of Figures

List of Tables

1 | Introduction

Network Flow and Packing Problems

Networks are ubiquitous in our everyday lives — be it transportation, communication, or social networks. In each of these networks, several *nodes* (crossings, peers, persons) are connected to each other via *edges* (streets, cables, friendship). In most cases, one is interested in some solution for moving entities (commodities, messages, information) from one place to another in this network. This solution — which we refer to as a *network flow* — is restricted by several constraints like lower bounds on the amount of entities that need to be moved or upper bounds on the amount of entities that can be moved via one single edge. On the other hand, the quality of a solution is measured by some objective function. Usually, one is interested in solutions that transport flow from one point in the network to another in a somewhat most cost-effective way. We refer to the problem of finding such a solution as the *minimum cost flow problem*. Stated as a linear program, this problem is given as follows:

$$\min \sum_{e \in E} c_e \cdot x_e \tag{1.1a}$$

$$\text{s.t.} \sum_{e \in \delta^-(v)} x_e - \sum_{e \in \delta^+(v)} x_e = -b_v \qquad \text{for all } v \in V, \tag{1.1b}$$

$$0 \leqslant x_e \leqslant u_e \qquad \text{for all } e \in E. \tag{1.1c}$$

Thereby, x_e denotes the amount of flow that is routed via edge e while c_e and u_e denote the cost and capacity of the corresponding edges, respectively. The value b_v denotes the supply of entities at each node v. We give more insights in this general formulation in the subsequent chapter.

Another popular set of problems is the class of *packing problems*. One famous representative for this kind of problem is the *knapsack problem*, in which one is interested in the best way of packing items with a specific profit and weight into a knapsack without exceeding a maximum weight. In its most general form, for some matrix $A \in \mathbb{R}_{\geqslant 0}^{m \times n}$ with non-negative entries and two vectors $c \in \mathbb{R}_{>0}^n$ and $b \in \mathbb{R}_{>0}^m$ with positive entries, a (fractional) packing problem can be stated as follows:

$$\max c^\mathsf{T} x \tag{1.2a}$$

$$\text{s.t.} \ Ax \leqslant b, \tag{1.2b}$$

$$x \geqslant 0. \tag{1.2c}$$

The knapsack problem complies with this form with the additional restriction that the variables are required to be integral. At a first glance, the minimum cost flow problem and a packing problem do not seem to have much in common. However, according to the well-known flow decomposition theorem, each minimum cost flow can be decomposed into flows on simple paths and cycles (cf. (Ahuja et al., 1993)). In other words, we can see each minimum cost flow as a "bundle" of "packed" flows on paths and cycles that do not violate the edge capacities. Hence, each flow problem can be seen as a packing problem in its core, so both problems are in fact strongly related to each other. This central observation will be underlined and exploited throughout this thesis.

Contributions

The minimum cost flow problem is one of most investigated problems in the field of discrete optimization. A large variety of combinatorial algorithms have emerged over the last decades that make the problem tractable both from a theoretical as well as a practical point of view. However, since these algorithms became more and more tailored to the inherent structure of the minimum cost flow problem, slight changes to this structure often make the usage of the corresponding algorithms or even their underlying ideas impossible to apply. An extension of the model by new constraints or a modification of the given constraints in (1.1) may influence the complexity and approximability of the problem significantly. Similarly, although dynamic programming schemes and approximation algorithms are known for the knapsack problem, additional constraints typically lead to much more difficult variants of the problem.

In this thesis, we investigate novel extensions to well-known network flow and packing problems. In particular, we are interested in results about the complexity and approximability of these problems and seek to find efficient combinatorial algorithms that exploit the underlying structure of the corresponding models, in contrast to highly generic simplex-type methods. Among others, this thesis addresses the following major issues:

- We investigate a *network improvement problem*, in which the capacities of the edges can be upgraded up to a specific amount by spending a separate upgrade budget. This problem is equivalent to the addition of an additional *budget constraint* of the form $\sum_{e \in E} b_e \cdot x_e \leqslant B$ to the formulation (1.1).

- We address an extended network flow problem in which the amount of flow entering an edge is not only bounded by the capacity constraints (1.1c), but must also satisfy constraints that restrict the distribution of flow among all outgoing edges of each node.

- We investigate the case in which the flow that leaves an edge does not necessarily equal the flow entering the edge, but is given by a increasing convex function depending on the entering flow. In other words, we study a variant of problem (1.1) in which constraint (1.1b) is generalized.

- In addition to these three network flow problems, we consider a variant of the traditional knapsack problem in which one is confronted with a set of *cardinality constraints* of the form $\sum_{i \in I_j} x_i \leqslant \mu_j$ for all sets I_j in a laminar family. We show a connection of this problem to the above network improvement problem.

- Finally, we investigate how a well-known framework to derive approximation algorithms for fractional packing problems can be generalized to a large class of network flow problems.

Outline of the Thesis

This thesis is divided into eight chapters. Besides this introductory chapter and a conclusion of the results in Chapter 8, the above mentioned issues are dealt with in the remaining six chapters as follows:

Chapter 2: This preliminary chapter provides the reader with the notational and conceptional preliminaries used throughout the thesis. In particular, we settle the basic notation and define the underlying model of computation. Moreover, we introduce the most common network flow problems that will be used and extended throughout the thesis and define the notion of approximation algorithms.

Chapter 3: In addition to the concepts introduced in Chapter 2, we introduce three generic frameworks that will be used throughout this thesis in Chapter 3. We combine two of these frameworks into a generalized packing framework that may be used to derive efficient approximation algorithms for fractional packing problems on finitely generated polyhedral cones. This generalized framework may in particular be applied to network flow problems and will be used in Chapter 4 and Chapter 6.

Chapter 4: In this chapter, we study an extension of the well-known minimum cost flow problem in which a second kind of costs is associated with each edge. The goal is to minimize the first kind of costs as in traditional minimum cost flows while the secondary costs of the flow must additionally fulfill a budget constraint. We present a specialized network simplex algorithm for this problem and provide both a weakly and a strongly polynomial-time algorithm. Moreover, we derive approximation schemes for the problem on general and on acyclic graphs.

Chapter 5: We consider two discrete variants of the problem investigated in Chapter 4. We show that both variants may be interpreted as network improvement problems, which yields several fields of applications. While both variants are hard to solve and approximate, in contrast to the original problem considered in Chapter 4, we show that the problem on extension-parallel graphs is strongly related to a novel variant of the bounded knapsack problem. By adapting results of the latter problem, we are able to derive efficient approximation algorithms for both variants.

Chapter 6: In this chapter, we present a generalization of the maximum flow problem in which each edge is assigned with a *flow ratio* that imposes an upper bound on the fraction of the total outgoing flow at the starting node that may be routed through the edge. While a flow decomposition can still be found efficiently for this problem, we prove that the problem becomes at least as hard to solve as any packing LP. On the other hand, we derive an efficient approximation scheme for the problem on acyclic graphs. For the case of series-parallel graphs, we provide two exact algorithms with strongly polynomial running time. Finally, we study the case of integral flows and show that the problem becomes hard to solve and approximate.

Chapter 7: As a third and last network flow problem, we give insights into the structural properties and the complexity of an extension of the generalized maximum flow problem in which the outflow of an edge is a strictly increasing convex function of its inflow. We show that the problem becomes hard to solve and approximate in this novel setting even in the fractional case. Nevertheless, we show that a flow decomposition similar to the one for traditional generalized flows is possible and present (exponential-time) exact algorithms for computing optimal flows on specific graph classes. We also identify a polynomially solvable special case of the problem and show that the problem is solvable in pseudo-polynomial time when restricting to integral flows on series-parallel graphs.

Credits

Most of the results in this thesis are based on joint work with Sven O. Krumke and Clemens Thielen. Parts of the results of Chapter 4 and Chapter 5 are published in (Holzhauser et al., 2016a). More of the findings presented in Chapter 4 (Holzhauser et al., 2015a, 2016b) as well as the subject matter covered in Chapter 6 are submitted for publication (Holzhauser et al., 2016c). Finally, parts of the results obtained in Chapter 7 are published in (Holzhauser et al., 2015b). I am extremely grateful to the co-authors for their valuable time and for allowing me to include our joint work in this thesis.

2 | Preliminaries

In this chapter, we introduce the basic definitions, on which the remainder of this thesis relies. In particular, we settle the basic notation and define the underlying model of computation. Moreover, we introduce the most common network flow problems that will be used and extended throughout the thesis and define the notion of approximation algorithms. In the very most cases, we comply with standard notations and definitions, so the familiar reader may skip parts of this chapter at will.

2.1 Basic Notation

Throughout this thesis, we denote by $\mathbb{R}$ ($\mathbb{Q}$, $\mathbb{Z}$, $\mathbb{N}$) the set of *real* (*rational, integral, natural*) numbers. We denote the corresponding subsets of $\mathbb{R}$ and $\mathbb{Q}$ that contain all non-negative numbers by $\mathbb{R}_{\geqslant 0}$ and $\mathbb{Q}_{\geqslant 0}$, respectively. Similarly, we write $\mathbb{R}_{>0}$ and $\mathbb{Q}_{>0}$, respectively, for the subsets of all positive numbers. The set $\mathbb{N}$ of natural numbers does not contain zero while $\mathbb{N}_{\geqslant 0}$ does. For two sets A and B, we use the notation $A \subseteq B$ ($A \supseteq B$) to denote that A is a subset (superset) of B. If A is a proper subset (superset) of B, we write $A \subset B$ ($A \supset B$). Moreover, we denote the union (intersection) of the two sets A and B by $A \cup B$ ($A \cap B$). Finally, we write $\emptyset$ to denote the empty set.

We use the notation $A \in B^{m \times n}$ to denote that A is a matrix with m rows and n columns, each of which contains elements from the set B. Moreover, we denote by A_{ij} the element in the i-th row and the j-th column of A and use $A_{i\cdot}$ and $A_{\cdot j}$ to denote the i-th row vector and the j-th column vector of A, respectively.

For a function $f \colon \mathbb{N}_{\geqslant 0} \to \mathbb{N}_{\geqslant 0}$, the set $\mathcal{O}(f(n))$ contains all functions $g \colon \mathbb{N}_{\geqslant 0} \to \mathbb{N}_{\geqslant 0}$ with the property that there are constants n_0 and c such that $g(n) \leqslant c \cdot f(n)$ for each $n \geqslant n_0$. Similarly, the set $\Omega(f(n))$ contains all functions $g \colon \mathbb{N}_{\geqslant 0} \to \mathbb{N}_{\geqslant 0}$ with the property that there are constants n_0 and c such that $g(n) \geqslant c \cdot f(n)$ for each $n \geqslant n_0$. The intersection of $\mathcal{O}(f(n))$ and $\Omega(f(n))$ is denoted by $\Theta(f(n)) := \mathcal{O}(f(n)) \cap \Omega(f(n))$.

Finally, we denote the *logarithm* of the number a to the basis b by $\log_b a$. Whenever the basis of the logarithm is omitted, it can be assumed to be 2 without loss of generality. We denote the natural logarithm of a by $\ln a$ (i.e., the logarithm to the basis e, where

e is Euler's number). Moreover, we let sgn: $\mathbb{R} \to \{-1, 0, 1\}$ denote the *sign function*, which returns -1, 0, or 1 depending on whether the given argument is negative, zero, or positive.

2.2 Theory of Computation

In this section, we give a short definition of the computational model that is used throughout this thesis and introduce the tools that will be used to measure the computational complexity of the upcoming problems. Again, we will thereby stick to basic definitions and notations such that the experienced reader may skip parts of this section at will. In-depth treatments of the upcoming topics can be found in (Garey and Johnson, 1979; Grötschel et al., 1993; Motwani and Raghavan, 1995; Papadimitriou, 1994; Blum et al., 1989).

Instance Encoding

Throughout this thesis, we assume that the instances of the problems that investigate are encoded by using a "reasonable" *encoding scheme* into a string over some alphabet Σ. The set of all strings over Σ will be denoted by Σ^*. The *encoding length* or *size* of a problem instance I, denoted by $|I|$, is then the length of such a string. Although the number of symbols in Σ will be of no great importance for our results (as long as $|\Sigma| \geqslant 2$), we will assume a *binary encoding* $\Sigma = \{0, 1\}$. As a result, an integer of value n will, e.g., have a size of $\lceil \log_2 |n| + 1 \rceil$ bits in any instance (Grötschel et al., 1993; Garey and Johnson, 1979).

Computational Models

The complexity results stated throughout this thesis are based on the *random access machine (RAM) model*[1234], which is an alternative to the classical *Turing machine* (Papadimitriou, 1994) that is capable of an infinite set of registers, each of which can store one integer of arbitrary size and sign. A random access machine supports a set of instructions such as direct and indirect addressing of registers, jumping and branching, comparisons, as well as arithmetic operations such as addition, subtraction, multiplication and division of numbers. More precisely, we will stick to the *unit-cost RAM*, in which each of these operations can be performed in constant time, independent of the size of the involved integers (in contrast to the *log-cost RAM*, which accounts for the size of the operands). This simplification leads to a "too powerful" model in comparison to the Turing machine since we can generate very large numbers too quickly using repeated multiplication. Nevertheless, an equivalence between the models (up to a logarithmic factor) holds in case that the encoding length of the involved integers is polynomially bounded by the encoding length of the instance (Motwani and Raghavan, 1995). Since the numbers that are used in the Chapters 3 – 6 fulfill this property, we can use the unit-cost RAM without loss of generality. The *(worst case) time complexity* or *(worst case) running time* of an algorithm is a function $f\colon \mathbb{N}_{\geqslant 0} \to \mathbb{N}_{\geqslant 0}$ such that $f(|I|)$ denotes the maximum number of instructions that are executed by the random access machine at an input of size $|I|$ (Grötschel et al., 1993).

Note that a random access machine is only capable of handling integral (or, more general, rational) numbers. In Chapter 7, however, we will be confronted with real numbers, which may have an infinite (explicit) representation. For this reason, we will stick to the *Blum-Shub-Smale model (BSS model)* in the corresponding chapter, which is basically a random access machine that is extended by the possibilities to store real rather than integral numbers in its registers and to evaluate rational functions on the register contents at unit cost (Blum et al., 1989; Blum, 1998).

Decision Problems

A *decision problem* Π is a problem that, for each instance I, either gives the answer YES or NO. That is, the problem Π can be seen as a subset of $\Sigma^* \times \{0, 1\}$ such that, for each instance I, either $(I, 0)$ or $(I, 1)$ is contained in Π (Grötschel et al., 1993). We say that a decision problem Π has a *time complexity of* $f(|I|)$ or *is solvable in* $f(|I|)$ *time* if there is an algorithm with time complexity $g(|I|)$ for $g \in \mathcal{O}(f(|I|))$ that decides whether or not some problem instance I of size $|I|$ is a YES-instance of Π. A decision problem Π is *solvable in (weakly) polynomial time* if it has a time complexity of $p(|I|)$ for some polynomial $p\colon \mathbb{N}_{\geqslant 0} \to \mathbb{N}_{\geqslant 0}$. Moreover, if Π has a time complexity of $p(m)$ and uses $q(|I|)$ space, where m denotes the number of integers that occur in the problem instance (regardless of the magnitude of the integers) and p and q are two polynomials, it is *solvable in strongly polynomial time*. Conversely, if Π is not solvable in polynomial time but has a time complexity that is polynomial in $|I|$ and the absolute value of the integers in I, we call it *solvable in pseudo-polynomial time* (Grötschel et al., 1993; Garey and Johnson, 1979).

The class $\mathcal{P}$ consists of all decision problems that are solvable in (weakly or strongly, but not pseudo-) polynomial time. The class $\mathcal{NP}$ consists of all decision problems that are *verifiable* in polynomial time, i.e., the set of problems Π for which there is a decision problem $\Pi' \in \mathcal{P}$ such that, for each Yes-instance I of Π and for some polynomial p', there is a *certificate* $x \in \Sigma^*$ with $|x| \leqslant p'(|I|)$ and $((I, x), 1) \in \Pi'$ (Grötschel et al., 1993).[1]

A decision problem Π is *polynomial-time reducible* to a decision problem Π' if there is a function t that transforms an instance I of Π into an instance $I' := t(I)$ of Π' in polynomial time such that I is a Yes-instance of Π if and only if I' is a Yes-instance of Π'. A decision problem Π is said to be $\mathcal{NP}$-*hard* if every problem in $\mathcal{NP}$ is polynomial-time reducible to Π. If, in addition, $\Pi \in \mathcal{NP}$, the decision problem is said to be $\mathcal{NP}$-*complete*. If the decision problem remains $\mathcal{NP}$-complete even if the numbers that occur in each problem instance I are polynomially bounded by $|I|$, the problem is *strongly* $\mathcal{NP}$-*complete*. Otherwise, we call the decision problem *weakly* $\mathcal{NP}$-*complete*. Unless $\mathcal{P} = \mathcal{NP}$, an $\mathcal{NP}$-complete decision problem is not solvable in polynomial time (Garey and Johnson, 1979).

Optimization Problems

In many problems that occur in practice, one is not only interested in the information whether or not a given instance is a Yes-instance (as in the case of decision problems) but wants to determine the somewhat best solution among all feasible solutions to the underlying problem with respect to some quality measurement. In an *optimization problem*, the aim is to determine an *optimal solution* x out of the set of *feasible solutions* X that maximizes (minimizes) some *objective function* $z: X \to \mathbb{R}$ in case of a *maximization problem* (*minimization problem*). The definition of time complexity as well as the notion of weakly, strongly, and pseudo-polynomial time solvability can then be applied to the case of an optimization problem without further ado. For the sake of convenience, we usually say that an optimization problem is *weakly or strongly* $\mathcal{NP}$-*hard* ($\mathcal{NP}$-*complete*) *to solve* if the corresponding decision problem that asks if there is a feasible solution $x \in X$ with $z(x) \geqslant k$ ($z(x) \leqslant k$) in case of a maximization problem (minimization problem) for some given value k is weakly or strongly $\mathcal{NP}$-complete to solve, respectively.

In a *k-criteria optimization problem*, the aim is to optimize a *set* of objective functions $z^{(j)}: X \to \mathbb{R}$ for $j \in \{1, \ldots, k\}$ over the set X of feasible solutions, each of which can either be a maximization or a minimization objective. The *objective space* Z is then given by $Z := \left\{ \left(z^{(1)}(x), \ldots, z^{(k)}(x) \right)^{\mathsf{T}} : x \in X \right\}$. The notion of an "optimal solution" becomes ambiguous in case of a k-criteria optimization problem if $k \geqslant 2$. Instead, we call a solution $x \in X$ *efficient* if, for every other solution $x' \neq x \in X$, there is an index $j \in \{1, \ldots, k\}$ such that $z^{(j)}(x') < z^{(j)}(x)$ ($z^{(j)}(x') > z^{(j)}(x)$) in case of a maximization (min-

---

[1] An alternative definition of the class $\mathcal{P}$ ($\mathcal{NP}$) is that it contains all problems that are solvable in polynomial time on a *deterministic* (*non-deterministic*) Turing machine (Garey and Johnson, 1979).

imization) objective. The set $\mathcal{P} := \left\{ \left(z^{(1)}(x), \ldots, z^{(k)}(x) \right)^\mathsf{T} : x \in X \text{ and } x \text{ is efficient} \right\} \subseteq Z$ is then called the *pareto frontier* of the optimization problem. The corresponding set of efficient solutions is sometimes called pareto frontier as well.

2.3 Graph Theory

Throughout this thesis, we consider *directed (multi-)graphs* (or simply *graphs*) $G = (V, E)$, which are induced by a finite *node set* V and a finite multiset E, called *edge set*, that contains ordered pairs of $V \times V$. The elements in V and E are referred to as *nodes* and *edges*, respectively. We denote the cardinalities of V and E by n and m, respectively. At some points, we restrict our considerations to *simple graphs* in which E is a set rather than a multiset. If the edge set contains two-element subsets of V rather than ordered pairs of $V \times V$, the graph is called a *undirected*.

For each edge $e = (v, w) \in E$, we call v the *starting node* and w the *end node* of e and say that e *heads from* v *to* w. Similarly, we call e both an *outgoing edge* of v and an *ingoing edge* of w or simply say that e *leaves* v and *reaches* w. Two nodes $v \in V$ and $w \in V$ are furthermore called *adjacent* if there is an edge $e \in E$ that heads from v to w or vice versa. Likewise, we call a node v and an edge e *incident* if v is the starting or the end node of e. Moreover, we call two edges e and e' *adjacent* if they are both incident to the same node v.

For each node $v \in V$, we denote by $\delta^+(v) := \{e \in E : e \text{ leaves } v\}$ and $\delta^-(v) := \{e \in E : e \text{ reaches } v\}$ the set of *outgoing edges* and *ingoing edges* of v, respectively. Accordingly, we call $|\delta^+(v)|$ and $|\delta^-(v)|$ the *out-degree* and *in-degree* of node v, respectively. For a set $V' \subseteq V$, we write $\delta^+(V')$ ($\delta^-(V')$) to denote the set of edges $e = (v, w) \in E$ with $v \in V'$ and $w \notin V'$ ($v \notin V'$ and $w \in V'$).

For a given graph $G = (V, E)$, we call each graph $H = (V', E')$ with $V' \subseteq V$ and $E' \subseteq E$ a *subgraph* of G. For a given subset $V' \subseteq V$ of the node set, we call $G[V']$ the *subgraph induced by* V', which consists of all nodes $v \in V'$ and all edges $e \in E \cap (V' \times V')$. Similarly, for a subset $E' \subseteq E$ of the edge set, the graph $G[E']$ denotes the *subgraph induced by* E', i.e., the graph with edge set E' and the node set V' that contains all end nodes of edges in E'.

A sequence $P := (e_1, \ldots, e_k)$ of edges in E in which the end node of e_i and the starting node of e_{i+1} coincide for each $i \in \{1, \ldots, k-1\}$ is called a *path (of length k)*. If v_0 denotes the starting node of e_1 and v_k the end node of e_k, then P is also referred to as a v_0-v_k-path. Moreover, we say that P *connects* v_0 *and* v_k and that v_k *is reachable from* v_0 *(via P)*. If all of the involved nodes are distinct, we call P a *simple path*. A path $C := (e_1, \ldots, e_k)$

in which the end node of e_k and the starting node of e_1 coincide is called a *circuit (of length k)*. Furthermore, if the paths $P_1 := (e_1, \ldots, e_{k-1})$ and $P_2 := (e_2, \ldots, e_k)$ are both simple, we call C a *simple cycle (of length k)simple cycle* or just *cycle*. We call a (simple) path, circuit, or cycle *undirected* if we obtain a corresponding (simple) path, circuit, or cycle by reverting the direction of one or more of the contained edges.

For two distinguished nodes $s, t \in V$, an *s-t-cut* (S, T) is a partition of the node set into two disjoint sets S and T such that $s \in S$ and $t \in T$. We denote by $\delta^+(S)$ ($\delta^-(S)$) the set of *forward edges* (*backward edges*) in the cut. Usually, we also use the set of forward edges $\delta^+(S)$ to refer to the s-t-cut (S, T).

We call a graph $G = (V, E)$ *connected* if, for each two nodes $v_1, v_2 \in V$, there is a (possibly undirected) v_1-v_2-path P in G. Accordingly, we call G *strongly connected* if there is both a directed path from v_1 to v_2 and a directed path from v_2 to v_1 for every pair of nodes $v_1, v_2 \in V$. Each maximal subgraph (with respect to the edge set) of G that is (strongly) connected is called a *(strongly) connected component*.

A graph $G = (V, E)$ with the property that its node set V can be partitioned into two sets V_1 and V_2 such that $E \subseteq (V_1 \times V_2) \cup (V_2 \times V_1)$ is called *bipartite*. A graph G is called *acyclic* if it does not contain cycles. A *topological sorting* of a graph $G = (V, E)$ with node set $V = \{v_1, \ldots, v_n\}$ is a sequence $(v_{i_1}, \ldots, v_{i_n})$ with $i_j \neq i_l$ for $j \neq l$ (i.e., a ordering of the node set) such that $i_j < i_l$ for each edge $e = (v_{i_j}, v_{i_l}) \in E$. As it is well-known, a graph is acyclic if and only if it has a topological sorting (cf., e.g., (Ahuja et al., 1993)).

A *tree* is a connected graph that does not contain undirected cycles. A graph G in which each connected component is a tree is called a *forest*. We call a tree $T = (V, E)$ *rooted* if there is some distinguished *root node* $r \in T$. In such a rooted tree T, we say that a node $v \in V$ is on *level* k if it is connected to the root node by an undirected path P_v of length k. Every node $w \neq v$ on this path P_v is called an *ancestor* of v while v itself is a *successor* of w. If, in addition, there is an edge from v to w or from w to v in T, we call v a *child node* of w and w the *parent node* of v. Each node v that has no children is called a *leaf node* or simply *leaf* while every other node of T (including the root node) is called an *inner node*. If every inner node of the tree has exactly two children, we call T a *binary* tree. Furthermore, if every node is reachable from the root node via a directed path, we call the tree an *out-tree*. Conversely, if the root node is reachable from every node via a directed path, we call the tree an *in-tree*. The subgraph of a tree $T = (V, E)$ that is induced by a node $v \in V$ and all of its successors is called a *subtree of T*. Finally, for a graph $G = (V, E)$, we call a tree T a *spanning tree of G* if it is a subgraph of G and contains all nodes in V.

Another important class of graphs that is used throughout this thesis is the class of *series-parallel graphs*. Each series-parallel graph $G = (V, E)$ is associated with a *source* $s \in V$ and a *sink* $t \in V$ and can be recursively defined as follows:

Single edge: Each graph that consists of a single edge $e = (s, t)$ is series-parallel with source s and sink t (denoted by $G = e$).

Parallel composition: For two series-parallel graphs G_1, G_2 with sources s_i and sinks t_i, $i \in \{1, 2\}$, the graph G that is obtained by identifying s_1 with s_2 and t_1 with t_2 is series-parallel with source $s_1 = s_2$ and sink $t_1 = t_2$ (denoted by $G = G_1 \mid G_2$).

Series composition: For two series-parallel graphs G_1, G_2 with sources s_i and sinks t_i, $i \in \{1, 2\}$, the graph G that is obtained by identifying t_1 with s_2 is series-parallel with source s_1 and sink t_2 (denoted by $G = G_1 \circ G_2$).

In particular, note that series-parallel graphs are acyclic and connected according to the above definition. If, additionally, in each series composition of G_1 and G_2 at least one of G_1 or G_2 consists of a single edge, the graph is called *extension-parallel*. A *decomposition tree* T of a series-parallel graph G is a binary tree in which the leaves correspond to single edges of G and each inner node v either corresponds to a series or a parallel composition of the two series-parallel graphs that are induced by the leaves of the two subtrees of v. Such a decomposition tree T of a series-parallel graph G with m edges and n nodes contains m leaves, $n - 2$ inner nodes that correspond to series compositions, and $m - n + 1$ inner nodes that correspond to parallel compositions. Moreover, such a decomposition tree can be constructed from a given series-parallel graph in $\mathcal{O}(m)$ time (cf. Valdes et al. (1982)). A series-parallel graph G and a corresponding decomposition tree T are depicted in Figure 2.1. Note that this graph is not extension-parallel while the subgraph that is induced by the nodes $\{s, v, w\}$ is.

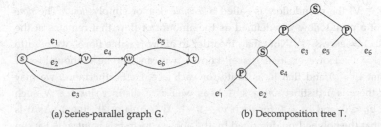

(a) Series-parallel graph G. (b) Decomposition tree T.

Figure 2.1: A series-parallel graph (left) and a possible decomposition tree (right). Inner nodes representing parallel compositions (series compositions) are denoted by the letter **P** (**S**).

Throughout this thesis, we assume that each graph $G = (V, E)$ is given in the *adjacency-list representation*, in which a list $\text{Adj}(v)$ is connected with each node $v \in V$ containing the outgoing edges $\delta^+(v)$ of v.

2.4 Network Flow Problems

In this section we give a short introduction to the field of *network flow problems*. In general, such problems aim at finding the "best" way to send some amount of a commodity from one point in a network to another one. Thereby, we will concentrate on the definitions and complexities of the four possibly most important network flow problems, namely the *shortest path problem*, the *maximum flow problem*, the *minimum cost flow problem*, and the *maximum generalized flow problem*. For an in-depth treatment of these topics, we refer the reader to Ahuja et al. (1993) and Wayne (1999).

In its most general form, each node $v \in V$ in a network flow problem is associated with an integral number b_v that represents the *supply* of the corresponding node. Moreover, each edge $e \in E$ has both a non-negative *lower capacity* $l_e \geqslant 0$ and a (possibly infinite) *upper capacity* or just *capacity* $u_e \geqslant l_e$. A *pseudoflow* is a function $x : E \to \mathbb{R}$ that assigns a value $x_e := x(e) \in [l_e, u_e]$ to each edge, which represents the amount of goods that are transported along e from the starting node to the end node of e. For a pseudoflow x, the *excess of a node* $v \in V$ if given as $\text{excess}_x(v) := \sum_{e \in \delta^-(v)} x_e - \sum_{e \in \delta^+(v)} x_e$ and describes the difference of the incoming amount of flow and the outgoing amount of flow at v. Similarly, the *imbalance of a node* $v \in V$ is given as $\text{excess}_x(v) + b_v$ and describes the deviation of the excess from the demand $-b_v$ of the node. If a pseudoflow x fulfills $\text{excess}_x(v) \geqslant -b_v$ at each $v \in V$, it is called a *preflow*. Moreover, if x fulfills the *flow conservation constraint* $\text{excess}_x(v) = -b_v$ at each node $v \in V$, the pseudoflow is called a *feasible flow* or simply *flow*. The *flow value* $\text{val}(x)$ of a pseudoflow x is defined as the amount of flow that remains at the sink, i.e., $\text{val}(x) := \text{excess}_v(t)$. Note that, in order to allow feasible flows, we require that $\sum_{v \in V} b_v = 0$. However, as it is well-known, we can assume without loss of generality that $l_e = 0$ and that u_e is finite for each $e \in E$. Furthermore, we may assume that there is a distinct *source* $s \in V$ as well as a distinct *sink* $t \in V$ such that $b_s > 0$, $b_t < 0$, and $b_v = 0$ for each $v \in V \setminus \{s, t\}$ (Ahuja et al., 1993). As it is well known that the polyhedron described by the above constraints is *integral*, we can assume without loss of generality that the optimal flow is integral as long as the input data is integral as well (Ahuja et al., 1993).

For a flow x in a network that is based on a graph $G = (V, E)$, the *residual network* $G(x)$ contains at most two edges for each $e = (v, w) \in E$ in the original graph G: Unless

$x_e = u_e$, there is some *forward edge* $e^{(1)}$ heading from v to w with a (positive) capacity of $u_{e^{(1)}} := u_e - x_e$. Moreover, unless $x_e = 0$, a *backward edge* $e^{(2)}$ heads from w to v with a (positive) capacity of $u_{e^{(2)}} := x_e$.

Shortest Paths

For a given graph with edge labels c_e for each $e \in E$, the *shortest path problem* in general aims at finding a (directed) path P between two nodes $v, w \in V$ with a minimum *length* $c(P) := \sum_{e \in P} c_e$ among all such paths. In the *single-source shortest path problem*, one seeks to find a shortest path from some node $v \in V$ to every other node $w \in V \setminus \{v\}$. Conversely, in the *all-pairs shortest path problem*, the task is to determine a shortest path between every pair of nodes in V. In particular, the shortest path problem is in fact a network flow problem since it can be seen as the problem of shipping one unit of a good from one point in the network to each of the $n - 1$ other points in the *cheapest* possible way. As we will see later, the problem can be seen as a special case of the more general minimum cost flow problem.

At present, the best bound for the single-source shortest path problem in a simple graph with non-negative lengths c_e is given by

$$SP(m, n, C) \in \mathcal{O} \left(\min \left\{ m + n \log n, m \log \log C, m + n\sqrt{C} \right\} \right),$$

with $C := \max_{e \in E} c_e$, where the corresponding bounds are due to Fredman and Tarjan (1987), Johnson (1981), and Ahuja et al. (1990), respectively. The best strongly polynomial time bound is consequently given by $SP(m, n) \in \mathcal{O}(m + n \log n)$. Moreover, in an acyclic graph, the single-source shortest path problem can be solved in $\mathcal{O}(m + n)$ time (Ahuja et al., 1993). Algorithms for the all-pairs shortest path problem will be investigated in Chapter 4.

Maximum Flows

In the *maximum flow problem*, the task is to determine a flow that sends the maximum possible amount of flow from the source s to the sink t of the network without violating any edge capacity. More precisely, the aim is to find a feasible flow x with maximum flow value $\text{val}(x)$ among all feasible flows or, equivalently, to determine the largest value $b_s = -b_t$ that allows a feasible flow x. Hence, we are able to leave out the flow conservation constraints for s and t and can maximize over $-b_t = \text{excess}_x(t)$

instead. Stated as a linear program, we then obtain the following formulation for the maximum flow problem:

$$\max \sum_{e \in \delta^-(t)} x_e - \sum_{e \in \delta^+(t)} x_e \tag{2.1a}$$

$$\text{s.t.} \sum_{e \in \delta^-(v)} x_e - \sum_{e \in \delta^+(v)} x_e = 0 \qquad \text{for all } v \in V \setminus \{s, t\}, \tag{2.1b}$$

$$0 \leqslant x_e \leqslant u_e \qquad \text{for all } e \in E. \tag{2.1c}$$

Note that we can transform each instance of the maximum flow problem on a multi-graph into an equivalent instance on a simple graph in linear time *without* increasing the number of nodes and edges by simply aggregating the capacities of all parallel edges between two nodes into the capacity of a new artificial edge and deleting the previous edges in $\mathcal{O}(m)$ time.

The maximum flow problem is probably the network flow problem with the longest history of steady improvements. The first algorithm for the maximum flow problem was introduced in 1956 by Ford and Fulkerson (1956) with a pseudo-polynomial running time of $\mathcal{O}(nmU)$ for $U := \max_{e \in E} u_e$. The underlying idea of repeatedly sending flow on s-t-paths with positive capacity in the residual network was later independently refined by Dinic (1970) and Edmonds and Karp (1972), resulting in strongly polynomial running times of $\mathcal{O}(n^2m)$ and $\mathcal{O}(nm^2)$, respectively. Another class of algorithms, called *push-relabel algorithms*, in which flow is augmented along single edges rather than full s-t-paths, was introduced by Karzanov (1974) and Gold-berg and Tarjan (1986). It resulted in running times of $\mathcal{O}(nm \log_{m/(n \log n)} n)$ and $\mathcal{O}(\min\{n^{2/3}, m^{1/2}\} \cdot m \log(n^2/m) \log U)$ due to King et al. (1994) and Goldberg and Rao (1998), respectively. In 2013, after there was no significant progress in the field of maximum flows for about 15 years, Orlin (2013) was able to give an affirmative answer to the long standing open question whether there is an algorithm with a running time in $\mathcal{O}(nm)$ by combining the ideas of King et al. (1994) and Goldberg and Rao (1998) with a new algorithm for sparse graphs. At present, the best bound for the maximum flow problem is given by

$$MF(m, n, U) \in \mathcal{O}\left(\min\left\{\min\{n^{2/3}, m^{1/2}\} \cdot m \log(n^2/m) \log U, nm\right\}\right)$$

due to Goldberg and Rao (1998) and Orlin (2013), respectively. The best strongly polynomial time bound is given by $MF(m, n) \in \mathcal{O}(nm)$ due to Orlin (2013).

Minimum Cost Flows

The *minimum cost flow problem* is the most fundamental network flow problem, particularly since it subsumes both the shortest path and the maximum flow problem and is strongly related to the maximum generalized flow problem that is described in the next subsection (Truemper, 1977). For given *edge costs* $c_e \in \mathbb{Z}$ for each $e \in E$, we seek to obtain a flow x that minimizes the total *flow costs* $c(x) = \sum_{e \in E} c_e \cdot x_e$ among all feasible flows. Note that we obtain the shortest path problem by setting $b_s := 1$ and $b_t := -1$ (if the flow is integral, which we can assume without loss of generality as described above) and the maximum flow problem by setting $c_e = -1$ for $e \in \delta^-(t)$ and $c_e = 0$ for $e \in E \setminus \delta^-(t)$.

Throughout this thesis, we usually restrict our considerations to the case that no flow value is prescribed, i.e., we drop the flow conservation constraints both for the source and the sink of the network (similar to the case of *minimum cost circulations*, cf. (Ahuja et al., 1993)). This assumption, however, does not constrain the capabilities of the model, which can be seen as follows: On the one hand, it can be shown that every minimum cost flow with a flow value of F decomposes into *some* flow with value F (which can be found by a maximum flow computation in $\mathcal{O}(nm)$ time as shown before) and a minimum cost circulation in the residual network of this flow. On the other hand, we can model a desired flow value of $F \geqslant 1$ by adding an artificial edge $\bar{e}$ heading from the original sink t to a new artificial sink t' with capacity $u_{\bar{e}} := F$ and cost $c_{\bar{e}} := -(mCU + 1)$. Since the absolute cost of any flow is bounded by mCU, the cost of a minimum cost flow in this transformed network is then smaller than $-(F-1)(mCU+1)$ if and only if it has a value of F and is minimum among all such flows. Note that this transformation is only connected with no loss of generality when talking about exact algorithms. For approximation algorithms (see Section 2.5), the transformation has an influence on the approximation guarantee of an algorithm. Nevertheless, as one usually needs to make assumptions in order to guarantee a non-negative or non-positive objective function value for a proper definition of an approximation guarantee, we restrict our considerations on the flow model described above.

Hence, we can formulate the minimum cost flow problem as a linear program as follows:

$$\min \sum_{e \in E} c_e \cdot x_e$$
$$\text{s.t.} \sum_{e \in \delta^-(v)} x_e - \sum_{e \in \delta^+(v)} x_e = 0 \qquad \text{for all } v \in V \setminus \{s, t\},$$
$$0 \leqslant x_e \leqslant u_e \qquad \text{for all } e \in E.$$

Note that the total flow costs are non-positive for each minimum cost flow in this setting since the all-zero flow is always feasible.

Similarly to the case of the maximum flow problem, the minimum cost flow problem has a long history of steady research and improvements. At present, the best bound $MCF(m, n, C, U)$ for the minimum cost flow problem is given by

$$MCF(m, n, C, U) \in \mathcal{O}(\min\{nm \cdot \log(n^2/m) \log(nC), nm \cdot \log \log U \log(nC),$$
$$m \log n \cdot (m + n \log n)\}).$$

These bounds are due to Goldberg and Tarjan (1987), Ahuja et al. (1992), and Orlin (1993), respectively. The best strongly polynomial time bound $MCF(m, n)$ is achieved by Orlin's *enhanced capacity scaling algorithm* (cf. Ahuja et al. (1993) and Orlin (1993)) and is given by $MCF(m, n) \in \mathcal{O}(m \log n \cdot (m + n \log n))$.

Note that each of these bounds only applies to *simple* graphs. Clearly, we can convert every multigraph into a simple graph by replacing each edge $e = (v, w)$ by two edges $e_1 = (v, v')$ and $e_2 = (v', w)$ for an artificial node v'. This transformation increases the number of edges from m to $2m$ and the number of nodes from n to $n + m$. However, in Chapter 4, we will see that we can avoid this transformation for the case of the enhanced capacity scaling algorithm, which in turn yields a time bound of $\mathcal{O}(m \log m \cdot (m + n \log n))$ for the problem on multigraphs.

Maximum Generalized Flows

The is an extension of the traditional maximum flow problem in which the implicit assumption of flow being conserved when it traverses an edge is dropped. Instead, the flow x_e that enters some edge e will have a value of $\gamma_e \cdot x_e$ for some *gain factor* $\gamma_e > 0$ when it leaves that edge. For different values of these gain factors, one can model effects like evaporation in a gas pipeline (if $\gamma_e < 1$) or money exchanges among different currencies (Wayne, 1999). Clearly, the resulting polyhedron is no longer integral, so we cannot assume a maximum generalized flow to be integral as well. The maximum generalized flow problem can be formulated as a linear program as follows:

$$\max \sum_{e \in \delta^-(t)} \gamma_e \cdot x_e - \sum_{e \in \delta^+(t)} x_e$$

$$\text{s.t.} \sum_{e \in \delta^-(v)} \gamma_e \cdot x_e - \sum_{e \in \delta^+(v)} x_e = 0 \qquad \text{for all } v \in V \setminus \{s, t\}$$

$$0 \leqslant x_e \leqslant u_e \qquad \text{for all } e \in E.$$

Note that the capacity of an edge bounds the flow that *enters* an edge (rather than the flow that *leaves* the edge). At present, the best bound for the maximum generalized flow problem is given by

$$MGF(m, n, B) \in \mathcal{O}\left(\min\left\{nm \cdot (m + n \log n) \log B, m^5\right\}\right)$$

if the capacities are integers between 1 and B and the gain factors are fractions of numbers between 1 and B. The first time bound is due to (Radzik, 2004) while the strongly polynomial time bound is due to Végh (2013), who recently showed that the generalized maximum flow problem can be solved in strongly polynomial time $MGF(m, n) \in \mathcal{O}(m^5)$.

2.5 Approximation Algorithms

For a maximization (minimization) problem Π with non-negative objective function z, an algorithm is called an *approximation algorithm with performance guarantee* $\alpha \in [1, \infty)$ or simply α-*approximation algorithm* if, for each instance I of Π, it computes a feasible solution x with objective value $z(x) \geqslant \frac{1}{\alpha} \cdot z(x^*)$ $(z(x) \leqslant \alpha \cdot z(x^*))$ in polynomial time, where x^* denotes an optimal solution of I.

An algorithm that receives an instance I of a maximization (minimization) problem Π and a real number $\varepsilon \in (0, 1)$ as its input is called a *polynomial-time approximation scheme (PTAS)* if, on input (I, ε), it computes a feasible solution x with objective value $z(x) \geqslant (1 - \varepsilon) \cdot z(x^*)$ $(z(x) \leqslant (1 + \varepsilon) \cdot z(x^*))$ with a running time that is polynomial in the encoding size $|I|$ of I. If this running time is additionally polynomial in $\frac{1}{\varepsilon}$, the algorithm is called a *fully polynomial-time approximation scheme (FPTAS)*. Similarly, for a k-criteria optimization problem Π with objective functions $z^{(j)}$ for $j \in \{1, \dots, k\}$, we call an algorithm a k-*criteria fully polynomial-time approximation scheme* (k-*criteria FPTAS*) if, on input (I, ε) with $I \in \Pi$ and $\varepsilon \in (0, 1)$, it computes a feasible solution x with objective value $z^{(j)}(x) \geqslant (1 - \varepsilon) \cdot z^{(j)}(x^*)$ for each maximization objective $z^{(j)}$ and $z^{(j)}(x) \leqslant (1 + \varepsilon) \cdot z^{(j)}(x^*)$ for each minimization objective $z^{(j)}$.

The notion of a k-criteria FPTAS is strongly related to the concept of ε-*approximate pareto frontiers*: For some instance I of a k-criteria optimization problem, the ε-approximate pareto frontier $\mathcal{P}(\varepsilon)$ is a subset of the feasibility set X such that, for each $x \in X$, there is a point $x_P \in \mathcal{P}(\varepsilon)$ that fulfills $z^{(j)}(x_P) \geqslant \frac{1}{1+\varepsilon} \cdot z^{(j)}(x)$ for each maximization objective and $z^{(j)}(x_P) \leqslant (1 + \varepsilon) \cdot z^{(j)}(x)$ for each minimization objective (cf. (Papadimitriou and Yannakakis, 2000)). Intuitively, the ε-approximate pareto frontier is a "sufficiently good" approximation of the real pareto frontier $\mathcal{P}$ with respect to the precision parameter ε. Note that, for $\varepsilon \in (0,1)$, the fact that $z^{(j)}(x_P) \geqslant \frac{1}{1+\varepsilon} \cdot z^{(j)}(x)$ also implies that $z^{(j)}(x_P) \geqslant (1-\varepsilon) \cdot z^{(j)}(x)$ as in the definition of an FPTAS above.

Finally, an optimization problem Π is said to be $\mathcal{NP}$-*hard to approximate* if the existence of an approximation algorithm with a specific performance guarantee α would imply that an $\mathcal{NP}$-hard decision problem is solvable in polynomial time.

3 | Fractional Packing and Parametric Search Frameworks

> The following chapter introduces three generic frameworks that will be used throughout this thesis. On the one hand, the focus will be set on the parametric search framework due to Megiddo (1979, 1983), which can often be used to solve parametric variants of known combinatorial problems in strongly polynomial time. On the other hand, we will review the fractional packing framework by Garg and Koenemann (2007), which yields generic fully polynomial-time approximation schemes (FPTASs) for problems that can be formulated as packing LPs[1]. In the second part of this chapter, we combine both approaches into a generalized packing framework that may be used to derive FPTASs for packing problems on finitely generated polyhedral cones. In particular, for a given oracle for this cone, the result extends to the case of cones that are generated by an exponential number of vectors. We show that we obtain FPTASs with varying time complexities for oracles with varying power. Finally, we show that this generalized packing framework, which will be used in Chapter 4 and 6, yields FPTASs for a large class of network flow problems in general.

3.1 Parametric Search

At several times throughout this thesis, we will make use of the *parametric search technique* due to Megiddo (1979), which often leads to faster algorithms for parametric problems in case that an algorithm is already known for the version without a parameter (Megiddo, 1979, 1983). We will briefly discuss the basic idea of this parametric search technique in the following before we come to a concrete application in Section 3.3.

Suppose that we have an algorithm $\mathcal{A}$ that solves a given optimization problem in $\mathcal{O}(T_{\mathcal{A}})$ time. Assume that we generalize the input of this problem from constant values to *linear parametric values* of the form $a_0 + \lambda \cdot a_1$ that linearly depend on some parameter λ, which yields a parametric version of the optimization problem. Moreover, assume that we want to find a specific value λ^* such that the optimum solution

---

1 A *packing LP* is a linear program of the form $\max\{c^\mathsf{T}x : Ax \leqslant b, x \geqslant 0\}$ where $c_j > 0$, $b_i > 0$, and $A_{ij} \geqslant 0$ for each $i \in \{1, \ldots, m\}$ and $j \in \{1, \ldots, n\}$.

to the problem with the input values $a_0 + \lambda^* \cdot a_1$ (which is then no longer parametric) achieves some specific objective value z^*. For example, in the *minimum ratio cycle problem* (also known as *minimum cost-to-time cycle problem*, cf. (Megiddo, 1979, 1983; Lawler, 2001)), one is interested in a cycle C^* in a graph $G = (V, E)$ that minimizes the value $\frac{\sum_{e \in C} c_e}{\sum_{e \in C} b_e}$ among all cycles C, where c_e and b_e are two kinds of costs connected with each $e \in E$ such that $\sum_{e \in C} b_e > 0$ for each cycle C. It is well-known that λ^* equals the objective function value of a minimum ratio cycle C^* if and only if the cost of a minimum cost cycle with edge costs $a_e - \lambda^* \cdot b_e$ for each $e \in E$ is zero (we will show a more general variant of this claim in Section 3.3).

Now assume that we have a *decision oracle* or *callback function* $\mathcal{B}$ that decides in $\mathcal{O}(T_\mathcal{B})$ time if a candidate value λ' for λ^* is smaller, larger, or equal to λ^*. In the example of the minimum ratio cycle problem, we only need to check the sign of a minimum mean cycle with (constant) edge costs $a_e - \lambda' \cdot b_e$, which can be done in $\mathcal{O}(nm)$ time (Karp, 1978).

Megiddo's (1979) parametric search technique now proceeds as follows: Let I denote an interval for which we know that it contains the optimal value λ^* (initially, we set $I := (-\infty, +\infty)$). We simulate the execution of the algorithm $\mathcal{A}$ with the linear parametric values $a_0 + \lambda \cdot a_1$ step by step, where λ is now handled as a *symbolic* variable. As long as we add two linear parametric values $a_0 + \lambda \cdot a_1$ and $b_0 + \lambda \cdot b_1$, the resulting value $(a_0 + b_0) + \lambda \cdot (a_1 + b_1)$ is linear parametric again. Similarly, if we subtract two linear parametric values or multiply a linear parametric value by a *constant*, the result is linear parametric again. In contrast, multiplications or divisions of two linear parametric values would destroy the linear parametric structure. In the following, we call an algorithm *strongly combinatorial* if it only uses the above three valid operations on its register contents and restrict our considerations on such types of algorithms in the following[2].

Now suppose that we encounter a branching in the execution path of $\mathcal{A}$ that depends on a comparison of two linear parametric values $a_0 + \lambda \cdot a_1$ and $b_0 + \lambda \cdot b_1$. By interpreting both values as linear functions, we get that there are two cases to consider for the result of the comparison: Either $a_1 = b_1$, in which case the two functions do not intersect and the result of the comparison does not depend on the value of λ, or $a_1 \neq b_1$, in which case the two functions intersect at exactly one point λ'. In the first case, we can easily resolve the comparison by comparing a_0 to b_0 and proceed the simulation of $\mathcal{A}$. In the second case, we need to distinguish the two cases whether λ' lies outside or inside of I. In the first case, the comparison, again, does not depend

---

2   In fact, most of the known combinatorial algorithms for network flow problems turn out to be strongly combinatorial as well. In our example, it can be easily seen that the minimum mean cycle algorithm by Karp (1978) fulfills this property.

on the value of λ (since one of the two functions is greater than the other one within the set I of relevant values of λ) and we can resolve it in constant time. In the other case, however, the result of the comparison depends on the value of λ. Nevertheless, we can use the decision oracle $\mathcal{B}$ in order to decide if $\lambda^* = \lambda'$ or whether $\lambda' < \lambda^*$ or $\lambda' > \lambda^*$. In the first case, we have found the optimal value of λ and are able to set $I := \{\lambda^*\}$, which allows us to resolve the present and each subsequent comparison. If $\lambda' < \lambda^*$, the candidate value λ' was too small, which implies that all of the relevant values for λ are *larger* than λ', i.e., we can update the interval I to $I \cap (\lambda', +\infty)$. The comparison becomes then independent from the value of λ over the set I and we can proceed the execution as above. Similarly, if $\lambda' > \lambda^*$, we update I to $I \cap (-\infty, \lambda')$ and proceed as well.

After at most $\mathcal{O}(T_{\mathcal{A}})$ steps, the simulation of $\mathcal{A}$ finishes with an optimum solution (which may still depend on λ) with objective value $z_0 + \lambda \cdot z_1$. Solving $z_0 + \lambda \cdot z_1 = z^*$ for λ, we get the desired value λ^* (which consequently lies in I). By inserting λ^* in the parametric description of the optimum solution, we in turn obtain the exact optimum solution.

In the worst-case, we need to evaluate the callback for each of the $\mathcal{O}(T_{\mathcal{A}})$ comparisons that may encounter in the course of the simulation, which yields a total running time for the procedure of $\mathcal{O}(T_{\mathcal{A}} \cdot T_{\mathcal{B}})$. In the above example, we use the minimum mean cycle algorithm both for the strongly combinatorial algorithm $\mathcal{A}$ and the callback function $\mathcal{B}$ until we end with the value λ^* for which the minimum mean cycle has zero mean costs. This cycle then yields the minimum ratio cycle. Since a minimum mean cycle can be computed in $\mathcal{O}(nm)$ time, we get an overall strongly polynomial running time of $\mathcal{O}(n^2 m^2)$. This running time, however, can be significantly improved by incorporating parallelization techniques (Megiddo, 1979, 1983). One such technique is given in the following lemma:

Lemma 3.1:
For a parameter λ and a corresponding callback function $\mathcal{B}$ running in $\mathcal{O}(T_{\mathcal{B}})$ time, a set of k independent comparisons between linear parametric values depending on λ can be resolved simultaneously in $\mathcal{O}(k + \log k \cdot T_{\mathcal{B}})$ time.

Proof: As shown above, each of the k comparisons can be resolved by evaluating the callback function once, which would imply a running time of $\mathcal{O}(k \cdot T_{\mathcal{B}})$. However, we can improve this running time if we consider all comparisons simultaneously rather than sequentially: In a first step, we calculate the candidate value λ_i for each comparison *without* yet evaluating the callback function. We then determine the median λ_j among all candidate values, which can be done in $\mathcal{O}(k)$ time according to Blum et al. (1972) (see also (Cormen et al., 2009) for further details on this algorithm). For this median, we evaluate the callback function, which allows us to discard half of the candidate values and determine the outcome of their corresponding comparisons. For the remaining set, we again compute the median and so on until we know the outcome of each comparison. Since the size of the set of candidate values is halved in every step, we only need to invoke the callback function $\mathcal{O}(\log k)$ times at an overhead of $\mathcal{O}(2k)$ for the determination of the medians and for maintaining the candidate values. This shows the claim. $\qquad\square$

The above lemma is in particular useful in the case of parallel algorithms that exploit a large number of processors (Megiddo, 1983). The following lemma (which will turn out to be useful in Chapter 4) illustrates this fact:

Lemma 3.2:
Let $G = (V, E)$ denote a multigraph with linear parametric edge length $l_e(\lambda)$ for each $e \in E$ and let $\mathcal{B}$ denote a callback function for the parameter λ running in $T_\mathcal{B}(m, n) \in \Omega\left(\frac{m}{\log m}\right)$ time. The graph G can be turned into a simple graph G' that only contains the shortest edge among all parallel edges between two nodes in $\mathcal{O}(\log m \cdot \log \log m \cdot T_\mathcal{B}(m, n))$ time.

Proof: Let $S := \left\{(v, w) \in V^2 : |\delta^+(v) \cap \delta^-(w)| \geqslant 2\right\}$ denote the set of all pairs of nodes with at least two parallel edges between them. In order to determine the simple graph G' with the desired properties, we need to evaluate the minimum length of all edges in $\delta^+(v) \cap \delta^-(w)$ for each $(v, w) \in S$. As shown in (Valiant, 1975), we can determine the minimum of k values in $\mathcal{O}(\log \log k)$ time when using $\mathcal{O}(k)$ processors. We simulate the computation of this algorithm for all pairs in S in parallel, which results in a total number of $\mathcal{O}\left(\sum_{(v,w)\in S} |\delta^+(v) \cap \delta^-(w)|\right) = \mathcal{O}(m)$ processors. In order to reduce the number of callback evaluations, we simulate these $\mathcal{O}(m)$ processors sequentially in a round-robin manner until each of them either finishes its computation or halts at the comparison of two linear parametric values, yielding $\mathcal{O}(m)$ candidate values for λ that need to be resolved using the callback for λ. As shown in Lemma 3.1, we can resolve all of these comparisons simultaneously in $\mathcal{O}(m + \log m \cdot T_\mathcal{B}(m, n)) = \mathcal{O}(\log m \cdot T_\mathcal{B}(m, n))$ time and continue the simulation of the processors. After $\mathcal{O}(\log \log m)$ iterations of the above procedure, each processor has finished its computation and the edge with minimum length is determined for each $(v, w) \in S$, which shows the claim. $\qquad\square$

3.2 Fractional Packing Framework

In the course of this thesis, we will make use of the fractional packing framework introduced by Garg and Koenemann (2007), which allows to obtain fully polynomial-time approximation schemes for general packing problems. In the following, we will briefly discuss the basic idea of this framework and present the essential results of Garg and Koenemann (2007) that will be used throughout this thesis.

The fractional packing framework can be best understood at the example of the traditional maximum flow problem. In order to obtain an FPTAS, we use the *path-based formulation* for this problem: Since every maximum flow can be decomposed into at most m flows on s-t-paths P in the set $\mathcal{P}$ of all s-t-paths (cf. (Ahuja et al., 1993)), we can state the maximum flow problem as

$$\max \sum_{P \in \mathcal{P}} x_P \tag{3.1a}$$

$$\text{s.t.} \sum_{P \in \mathcal{P}: e \in P} x_P \leqslant u_e \qquad \text{for all } e \in E, \tag{3.1b}$$

$$x_P \geqslant 0 \qquad \text{for all } P \in \mathcal{P}, \tag{3.1c}$$

where x_P denotes the amount of flow that is sent on the path $P \in \mathcal{P}$. Consequently, for each edge $e \in E$, the sum of the flows on all paths that use e is bounded by the capacity u_e of the edge. Note that this formulation is in fact a packing LP, in contrast to the standard formulation of the maximum flow problem that was introduced in Section 2.4. The dual formulation of the primal problem (3.1) is given as follows:

$$\min \sum_{e \in E} y_e \cdot u_e$$

$$\text{s.t.} \sum_{e \in P} y_e \geqslant 1 \qquad \text{for all } P \in \mathcal{P},$$

$$y_e \geqslant 0 \qquad \text{for all } e \in E.$$

Although both the primal and the dual formulation of the problem are of exponential size in general, the fractional packing framework of Garg and Koenemann (2007) allows us to obtain an FPTAS for the maximum flow problem by using these formulations only implicitly, which will be shown in the following.

Suppose that we want to find an ε-approximate solution for the maximum flow problem with $\varepsilon \in (0, 1)$ and let $\varepsilon' := \frac{\varepsilon}{2}$. The procedure described in (Garg and Koenemann, 2007) starts with the feasible primal solution $x = 0$ and the infeasible dual solution given by $y_e := \frac{\delta}{u_e} > 0$ for each $e \in E$ with $\delta := \frac{(1+\varepsilon')}{((1+\varepsilon')m)^{\frac{1}{\varepsilon'}}}$. In each step of the algorithm, the *most violated dual constraint* (i.e., the constraint in the dual formulation with the largest relative deviation of the left-hand side value to the right-hand side value) that corresponds to some path $P \in \mathcal{P}$ is determined based on the current dual solution y. Although there are exponentially many constraints in the dual formulation, we can find the most violated constraint in $\mathcal{O}(SP(m, n))$ time by computing a shortest s-t-path with the edge lengths y_e for each $e \in E$. We then increase x_P for the shortest path P by the maximum value $u_P := \min_{e \in E} u_e$ (i.e., we send u_P units of flow on P) that does not violate any capacity constraint of the edges on P *without* considering the flow that has been sent in previous iterations, which will most likely make the primal solution infeasible. At the same time, each variable y_e for $e \in P$ will be multiplied by a factor of $\left(1 + \varepsilon' \cdot \frac{u_P}{u_e}\right)$. Intuitively, the "congested" edges will get "longer" over time and will, thus, be used less likely in future iterations, which somehow balances the flow among all paths.

The algorithm stops as soon as the dual solution fulfills $\sum_{e \in E} u_e \cdot y_e \geq 1$. As noted above, the primal solution will most likely be infeasible since, in each iteration, flow is added regardless of the existing flow in the network. However, Garg and Koenemann (2007) show that we obtain a feasible primal solution by scaling down the solution x by $\log_{1+\varepsilon'} \frac{1+\varepsilon'}{\delta}$ and that this solution is within a factor $(1 - 2\varepsilon') = (1 - \varepsilon)$ of the optimal solution. Moreover, they prove that the described procedure terminates within $\frac{1}{\varepsilon'} \cdot m \cdot (1 + \log_{1+\varepsilon'} m) = \mathcal{O}\left(\frac{1}{\varepsilon^2} \cdot m \log m\right)$ iterations. Subsequently, Garg and Koenemann (2007) generalized this idea from the maximum flow problem to the case of general packing problems of the form $\max\{c^\mathsf{T} x : Ax \leq b, x \geq 0\}$ for a cost vector c with positive entries, a right-hand side vector b with positive entries and a matrix A with non-negative entries. If N denotes the total number of non-zero entries in A and m the number of rows in A, the authors show that this general problem can be solved in $\mathcal{O}\left(\frac{1}{\varepsilon^2} \cdot m \log m \cdot N\right)$ time. We refer to (Garg and Koenemann, 2007; Fleischer and Wayne, 2002) for further details on the general procedure.

3.3 A Generalized Framework

In this section, we generalize the result of Garg and Koenemann (2007) to the case of packing problems on polyhedral cones. In order to do so, we incorporate Megiddo's parametric search technique that was described in Section 3.1 into the fractional packing framework considered in Section 3.2. We will also give insights into possible applications of this generalized framework.

In the following, let $S := \{x^{(1)}, \ldots, x^{(k)}\}$ denote a finite set of k non-negative n-dimensional vectors $x^{(l)} \in \mathbb{R}^n_{\geq 0}$ with $x^{(l)} \neq 0$. The cone that is spanned by these vectors is given by

$$C := \left\{ x \in \mathbb{R}^n : x = \sum_{l=1}^{k} \alpha_l \cdot x^{(l)} \text{ with } \alpha_l \geq 0 \text{ for all } l \in \{1, \ldots, k\} \right\}. \tag{3.2}$$

The main result of this section is that we are able to derive an FPTAS for the problem to maximize a linear function over the cone C subject to a set of packing constraints under specific assumptions that will be investigated in the following. As we will see, this extended framework is especially useful for the case of such network flow problems for which some kind of flow decomposition theorem is known.

Note that we do *neither* require the set S to be of polynomial size *nor* assume the set S or the cone C to be given explicitly. Instead, as it is common when dealing with implicitly given polyhedra, we only assume the cone to be *well-described*, which implies that it has an encoding length of at least $n + 1$ (cf. (Grötschel et al., 1993) for further details). Moreover, we make decisions over S and C via a given *oracle* $\mathcal{A}$ (to be specified later) that yields information about S based on a given cost vector d. In particular, since we are interested in polynomial-time approximation schemes, we assume the oracle to run in polynomial time depending on the dimension n of the ground set and the encoding length $\mathcal{O}(\sum_{j=1}^{n} \log d_i)$ of the cost vector d. Note that, based on this assumption, a vector $x \in C$ returned by the oracle has an encoding length that is polynomially bounded by n and the encoding length of d as well. Moreover, we assume that the running-time T_A of such an oracle $\mathcal{A}$ fulfills $T_A \in \Omega(n)$ since it would not be able to investigate each component of d or return a vector $x \in C$ otherwise.

In the following, let $A \in \mathbb{R}^{m \times n}_{\geq 0}$ denote a constraint matrix with non-negative entries, $b \in \mathbb{R}^m_{>0}$ a positive right-hand side vector, and $c \in \mathbb{R}^n$ the cost vector with arbitrary sign. Without loss of generality, we assume that at least one entry in each row and each column of A is positive. Moreover, we define N to be the number of non-zero entries contained in the matrix A.

As described above, the problem that we want to approximate is given as follows:

$$\max c^T x \tag{3.3a}$$
$$\text{s.t. } Ax \leq b, \tag{3.3b}$$
$$x \in C. \tag{3.3c}$$

We summarize the task of the generalized packing framework as follows:

Definition 3.3 (Generalized Packing Framework (GPF)):

INSTANCE: A cost vector $c > 0$, right-hand side vector $b > 0$, constraint matrix $A \geqslant 0$ as well as an implicitly given set S and an oracle $\mathcal{A}$ for S.

TASK: Return an FPTAS for the problem (3.3).

$\triangleleft$

Using the definition of the cone C based on equation (3.2), we obtain the following equivalent formulation of the problem (3.3):

$$\max c^T \sum_{l=1}^{k} \alpha_l \cdot x^{(l)}$$

$$\text{s.t. } A \left(\sum_{l=1}^{k} \alpha_l \cdot x^{(l)} \right) \leqslant b,$$

$$\alpha_l \geqslant 0 \qquad \qquad \text{for all } l \in \{1, \dots, k\}.$$

In particular, we replaced the original variables x by the weight vector α and, in doing so, incorporated the constraints of the cone. As noted above, this formulation might be of exponential size. However, in the following, we will never need to state it explicitly but will derive results based on its implicit structure. By rearranging the objective function and the packing constraints, we obtain the following equivalent formulation of the problem:

$$\max \sum_{l=1}^{k} \alpha_l \cdot \left(c^T x^{(l)} \right) \tag{3.4a}$$

$$\text{s.t. } \sum_{l=1}^{k} \alpha_l \cdot \left(A_i . x^{(l)} \right) \leqslant b_i \qquad \text{for all } i \in \{1, \dots, m\}, \tag{3.4b}$$

$$\alpha_l \geqslant 0 \qquad \text{for all } l \in \{1, \dots, k\}. \tag{3.4c}$$

Clearly, we can neglect vectors $x^{(l)}$ for which $c^T x^{(l)} \leqslant 0$ since, without loss of generality, it holds that $\alpha_l = 0$ for each such l in an optimal solution. Hence, in the following, we restrict our considerations on vectors $x^{(l)}$ with $c^T x^{(l)} > 0$ such that the primal problem (3.4) is in fact a packing LP (again, possibly of exponential size). The dual formulation of this problem is given as follows:

$$\min \sum_{i=1}^{m} y_i \cdot b_i \tag{3.5a}$$

$$\text{s.t. } \sum_{i=1}^{m} y_i \cdot \left(A_i . x^{(l)} \right) \geqslant c^T x^{(l)} \qquad \text{for all } l \in \{1, \dots, k\} \text{ with } c^T x^{(l)} > 0, \tag{3.5b}$$

$$y_i \geqslant 0 \qquad \text{for all } i \in \{1, \dots, m\}. \tag{3.5c}$$

As it was shown in Section 3.2, we can apply the fractional packing framework of Garg and Koenemann (2007) to the original problem (3.3) *provided* we are able to determine the most violated dual constraint in equation (3.5b) efficiently. Hence, given a dual solution $y > 0$, we need to be able to solve the following subproblem in polynomial time:

$$\min_{\substack{l \in \{1,\ldots,k\} \\ c^T x^{(l)} > 0}} \frac{\sum_{i=1}^m y_i \cdot \left(A_i. x^{(l)}\right)}{c^T x^{(l)}} = \min_{\substack{l \in \{1,\ldots,k\} \\ c^T x^{(l)} > 0}} \frac{\sum_{i=1}^m y_i \cdot \sum_{j=1}^n A_{ij} \cdot x_j^{(l)}}{c^T x^{(l)}}$$

$$= \min_{\substack{l \in \{1,\ldots,k\} \\ c^T x^{(l)} > 0}} \frac{\sum_{j=1}^n x_j^{(l)} \cdot \sum_{i=1}^m y_i \cdot A_{ij}}{c^T x^{(l)}}.$$

For $a_j := \sum_{i=1}^m y_i \cdot A_{ij}$ for $j \in \{1,\ldots,n\}$ and $a = (a_1,\ldots,a_n)^T$, this subproblem reduces to

$$\min_{\substack{l \in \{1,\ldots,k\} \\ c^T x^{(l)} > 0}} \frac{a^T x^{(l)}}{c^T x^{(l)}}. \tag{3.6}$$

Note that $a_j > 0$ for each $j \in \{1,\ldots,n\}$ since $y_i > 0$ for each $i \in \{1,\ldots,m\}$ throughout the procedure of Garg and Koenemann (2007) and since the matrix A has at least one positive and no negative entry in each row as assumed above. Since $x^{(l)} \neq 0$ and $x^{(l)} \in \mathbb{R}^n_{\geq 0}$ for each $l \in \{1,\ldots,k\}$, this also yields that $a^T x^{(l)} > 0$, so the minimum in equation (3.6) is strictly positive.

Clearly, if the vectors in S are given explicitly in the instance of GPF, we immediately obtain an FPTAS for the original problem (3.3) using the arguments given in Section 3.2. In the following, we discuss three cases in which we are still able to solve this subproblem efficiently even if we can access the set S and the cone C only via an oracle.

3.3.1 Minimizing Oracles

First consider the case that we have a *minimizing oracle* over S, which is defined as follows:

Definition 3.4 (Minimizing Oracle):
For a given vector $d \in \mathbb{R}^n$, a *minimizing oracle* for the set S returns a vector $x^{(l^*)} \in S$ that minimizes $d^T x^{(l)}$ among all vectors $x^{(l)} \in S$. ◁

For a given minimizing oracle for the set S, we are able to solve the following special case efficiently:

Theorem 3.5:
Suppose that $c^T x^{(l)} = \bar{c}$ for all $x^{(l)} \in S$ and some constant $\bar{c} > 0$. Given a minimizing oracle $\mathcal{A}$ for S running in T_A time, there is an FPTAS for the problem (3.3) with a running time in $\mathcal{O}\left(\frac{1}{\varepsilon^2} \cdot m \log m \cdot (N + T_A)\right)$.

Proof: Since $c^T x^{(l)} = \bar{c}$ for each $x^{(l)} \in S$, the subproblem given in equation (3.6) reduces to the problem of finding a vector $x^{(l)}$ with minimum cost $\left(\frac{1}{\bar{c}} \cdot a\right)^T x^{(l)}$ among all vectors in S. Using the minimizing oracle, we can compute a minimizer for (3.6) in $\mathcal{O}(T_A)$ time based on the cost vector $d := \frac{1}{\bar{c}} \cdot a$ (which can be built in $\mathcal{O}(N)$ time). Hence, we are able to determine the most violated dual constraint of (3.5) in $\mathcal{O}(N + T_A)$ time, so the claim follows immediately from the arguments outlined in Section 3.2. $\qquad\square$

Example 3.6:
In the *budget-constrained maximum flow problem*, the aim is to determine a flow with maximum value in an s-t-network that is additionally restricted by a *budget-constraint* of the form $\sum_{e \in E} b_e \cdot x_e \leqslant B$ for positive integers $b_e \in \mathbb{N}$ for each $e \in E$ and $B \in \mathbb{N}$ (cf. (Ahuja and Orlin, 1995)). Without loss of generality, since each budget-constrained maximum flow x is also a traditional s-t-flow and since flows on cycles do not contribute to the flow value, it holds that x can be decomposed into $m' \leqslant m$ flows $\bar{x}^{(j)}$ on s-t-paths P_j such that $x = \sum_{j=1}^{m'} \bar{x}^{(j)}$. Hence, if $x^{(l)}$ denotes the flow with unit flow value on some path P_l in the set of s-t-paths $\{P_1, \ldots, P_k\}$, it holds that each (budget-constrained) maximum flow x is contained in the cone C that is generated by the vectors in the set $S := \{x^{(l)} : l \in \{1, \ldots, k\}\}$. Hence, we can formulate the budget-constrained maximum flow problem as follows:

$$\max \sum_{e \in E} c_e \cdot x_e \tag{3.7a}$$

$$\text{s.t.} \sum_{e \in E} b_e \cdot x_e \leqslant B, \tag{3.7b}$$

$$x_e \leqslant u_e \qquad \text{for each } e \in E, \tag{3.7c}$$

$$x \in C, \tag{3.7d}$$

where $c_e = 1$ if $e \in \delta^-(t)$, and $c_e = 0$, else. In particular, it holds that $c^T x^{(l)} = 1$ for each $x^{(l)} \in S$ since each s-t-path contributes equally to the value of the flow. In order to apply Theorem 3.5, we need to show that there is a minimizing oracle for S, i.e., that we can determine a vector $x^{(l)}$ that minimizes $d^T x^{(l)}$ for a given cost vector d. However, this reduces to the determination of a shortest s-t-path with respect to the edge lengths d. According to Theorem 3.5, we, thus, get that there is an FPTAS for the budget-constrained maximum flow problem running in $\mathcal{O}\left(\frac{1}{\varepsilon^2} \cdot m \log m \cdot \text{SP}(m, n)\right)$ time since the number N of non-zero entries in the constraint matrix is bounded by $2m \in \mathcal{O}(SP(m,n))$. Note that this running time is still obtained even if we add up to $\mathcal{O}(m)$ different budget-constraints.

We want to stress that our framework allows to stick to the commonly used edge-based formulation of the problem, in which there is a linear number of variables defining the flow on single edges. In contrast, one is required to use the path-based formulation of the problem when using the original framework of Garg and Koenemann (2007) (as it was done in the problem (3.1) on page 23): The flow conservation constraints, which define the "shape" of a feasible flow, cannot be directly used in a formulation as a packing problem. These constraints, however, are now modeled by the containment in the cone C. $\triangleleft$

3.3.2 Sign Oracles

The result of Theorem 3.5 yields a fast FPTAS for the problem (3.3) but relies both on a severe restriction on the vectors in S and the existence of a rather powerful oracle. We will now relax these assumptions and focus on the more general case in which the costs $c^T x^{(l)}$ of the vectors in S differ. As it turns out, it suffices to assume a weaker oracle to be given for the set S, which can be defined as follows:

Definition 3.7 (Sign Oracle):
For a given vector $d \in \mathbb{R}^n$, a *sign oracle* for the set S returns a vector $x^{(l)} \in S$ with $\operatorname{sgn} d^T x^{(l)} = \operatorname{sgn} d^T x^{(l^*)}$, where $x^{(l^*)}$ minimizes $d^T x^{(i)}$ among all vectors in S. $\triangleleft$

Note that any minimizing oracle also implies a sign oracle for the set S, so the results of this subsection also apply to the case of a minimizing oracle. We now show that we can incorporate a strongly combinatorial sign oracle into Megiddo's parametric search framework in order to obtain an FPTAS for the problem (3.3) that does not rely on any further restrictions.

Let λ^* denote the value of the minimum in equation (3.6) and, for $\lambda \in \mathbb{R}$, let $d(\lambda) := a - \lambda c$ and $D(\lambda) := \min\{d(\lambda)^T x^{(l)} : l \in \{1, \ldots, k\} \text{ with } c^T x^{(l)} > 0\}$. Similar to the minimum ratio cycle problem that was considered in Section 3.1 (Lawler, 2001; Megiddo, 1979, 1983), we get the following characterization of the relation between the sign of $D(\lambda)$ and the sign of $\lambda^* - \lambda$:

Lemma 3.8:
For some given value of $\lambda \in \mathbb{R}$, it holds that $\operatorname{sgn}(D(\lambda)) = \operatorname{sgn}(\lambda^* - \lambda)$.

Proof: Let $L := \{l \in \{1, \ldots, k\} : c^T x^{(l)} > 0\}$. First, consider the case that $D(\lambda) > 0$. The claim follows by simple arguments:

$$
\begin{aligned}
D(\lambda) > 0 &\iff d(\lambda)^T x^{(l)} > 0 && \text{for all } l \in L \\
&\iff (a - \lambda c)^T x^{(l)} > 0 && \text{for all } l \in L \\
&\iff \frac{a^T x^{(l)}}{c^T x^{(l)}} > \lambda && \text{for all } l \in L \\
&\iff \lambda^* > \lambda.
\end{aligned}
$$

Conversely, if $D(\lambda) < 0$, we get the following equivalences by similar arguments:

$$
\begin{aligned}
D(\lambda) < 0 &\iff d(\lambda)^T x^{(l)} < 0 && \text{for some } l \in L \\
&\iff (a - \lambda c)^T x^{(l)} < 0 && \text{for some } l \in L \\
&\iff \frac{a^T x^{(l)}}{c^T x^{(l)}} < \lambda && \text{for some } l \in L \\
&\iff \lambda^* < \lambda.
\end{aligned}
$$

Finally, in the remaining case $D(\lambda) = 0$, it follows by continuity that $\lambda^* = \lambda$, which shows the claim. $\qquad\qquad\square$

According to Lemma 3.8, we can decide about the direction of the deviation between some candidate value λ and the desired value λ^* if we are able to determine the sign of $D(\lambda)$. Clearly, this sounds like a suitable task for a sign oracle for S. However, the value $D(\lambda)$ is defined to be the minimum of $d(\lambda)^T x^{(l)}$ among all vectors $x^{(l)}$ that *additionally* fulfill $c^T x^{(l)} > 0$ whereas the sign oracle is not required to consider such vectors only according to Definition 3.7. Nevertheless, as it will be shown in the following lemma, we can neglect this additional restriction when evaluating the sign oracle:

Lemma 3.9:
For any positive value of λ, it holds that $\text{sgn}(D(\lambda)) = \text{sgn}(d(\lambda)^T x^{(l)})$ where $x^{(l)}$ is the vector returned by a sign oracle for S.

Proof: First, consider the case that $\text{sgn}(d(\lambda)^T x^{(l)}) = -1$, i.e., that $d(\lambda)^T x^{(l)} < 0$. Inserting the definition of $d(\lambda)$, we get that $(a - \lambda c)^T x^{(l)} = a^T x^{(l)} - \lambda \cdot c^T x^{(l)} < 0$. Since both $a^T x^{(l)} > 0$ and $\lambda > 0$, it must hold that $c^T x^{(l)} > 0$ as well. Thus, we can conclude that $D(\lambda) \leqslant d(\lambda)^T x^{(l)} < 0$.

Now consider the case that $\text{sgn}(d(\lambda)^T x^{(l)}) = 0$. According to Definition 3.7, it holds that there are no vectors $x^{(i)} \in S$ with $d(\lambda)^T x^{(i)} < 0$, so $D(\lambda) \geqslant 0$. As in the previous case, we get that $(a - \lambda c)^T x^{(l)} = a^T x^{(l)} - \lambda \cdot c^T x^{(l)} = 0$ if and only if $c^T x^{(l)} > 0$ since both $a^T x^{(l)} > 0$ and $\lambda > 0$. Hence, we also get that $D(\lambda) \leqslant d(\lambda)^T x^{(l)} = 0$, so $D(\lambda) = 0$.

Finally, if $\mathrm{sgn}(d(\lambda)^T x^{(1)}) = 1$, there are no vectors $x^{(i)} \in S$ with $d(\lambda)^T x^{(i)} \leqslant 0$. Thus, it also holds that $D(\lambda) > 0$, which shows the claim. $\qquad\square$

Theorem 3.10:
Given a strongly combinatorial and strongly polynomial-time sign oracle $\mathcal{A}$ for the set S running in $T_{\mathcal{A}}$ time, there is an FPTAS for the problem (3.3) with a running time in $\mathcal{O}\left(\frac{1}{\varepsilon^2} \cdot m \log m \cdot (N + T_{\mathcal{A}}^2)\right)$.

Proof: Lemma 3.8 and Lemma 3.9 imply that λ^* is the unique value for λ for which the sign oracle returns a vector $x^{(1)} \in S$ with $d(\lambda)^T x^{(1)} = 0$. In particular, the returned vector $x^{(1)}$ is a minimizer for (3.6). Hence, since the values a_i can be computed in $\mathcal{O}(N)$ time, we are done if we are able to determine such a vector $x^{(1)}$ in $\mathcal{O}(T_{\mathcal{A}}^2)$ time.

Let $d(\lambda)$ be defined as above, where λ is now treated as a *symbolic* value that is contained in the interval I. Initially, we set I to $(0, +\infty)$ since the optimal value λ^* is known to be strictly positive (cf. equation (3.6)). Note that the costs $d(\lambda)_i = a_i - \lambda \cdot c_i$ fulfill the linear parametric value property. We simulate the execution of the sign oracle $\mathcal{A}$ at input $d(\lambda)$ using Megiddo's parametric search technique as described in Section 3.1. The underlying idea is to "direct" the control flow during the execution of $\mathcal{A}$ in a way such that it eventually returns the desired vector that minimizes (3.6).

Whenever we need to resolve a comparison between two linear parametric values that intersect at some point $\lambda' \in I$, we call the sign oracle itself with the cost vector $d := d(\lambda')$ in order to determine the sign of $D(\lambda')$. If $D(\lambda') = 0$, we found a minimizer for equation (3.6) and are done. If $D(\lambda') < 0$ $(D(\lambda') > 0)$, the candidate value λ' for λ^* was too large (too small) according to Lemma 3.8 and Lemma 3.9 such that we can update the interval I to $I \cap (-\infty, \lambda')$ $(I \cap (\lambda', +\infty))$, resolve the comparison, and continue the simulation of the oracle algorithm. After $\mathcal{O}(T_{\mathcal{A}})$ steps, the simulation terminates and returns a vector $x^{(1)} \in S$ that fulfills $d(\lambda^*)^T x^{(1)} = 0$ for the desired value $\lambda^* \in I$. Hence, this vector yields the most violated constraint in (3.5b). Since the described simulation runs in $\mathcal{O}(T_{\mathcal{A}}^2)$ time, we obtain an FPTAS with the claimed running time according to the arguments outlined in Section 3.2. $\qquad\square$

Note that we actually still obtain a polynomial running-time of the above algorithm even if we do not assume the sign oracle to run in *strongly* polynomial time but only to run in *weakly* polynomial time. However, the running-time of the resulting algorithm might exceed the stated time bound since the candidate values λ' that determine the input to the callback oracle are rational numbers whose representation might involve exponential-size numbers of the form $M^{T_{\mathcal{A}}}$ for some M with polynomial encoding length. Although the running-time of a weakly polynomial-time oracle algorithm depends only logarithmically on the size of these numbers, the running-time might still increase by a large (polynomial) factor.

Since any minimizing oracle also yields a sign oracle, we immediately get the following corollary from the results of Theorem 3.10:

Corollary 3.11:
Given a strongly combinatorial and strongly polynomial-time minimizing oracle $\mathcal{A}$ for S, there is an FPTAS for the problem (3.3) running in $\mathcal{O}\left(\frac{1}{\varepsilon^2} \cdot m \log m \cdot \left(N + T_{\mathcal{A}}^2\right)\right)$ time.

Example 3.12:
In the *budget-constrained minimum cost flow problem*, the aim is to determine a minimum cost flow subject to a budget constraint of the form $\sum_{e \in E} b_e \cdot x_e \leq B$, equivalently to the budget-constrained maximum flow problem that was studied in Example 3.6. When considering the circulation based version of the problem in which $\text{excess}_x(v) = 0$ for each $v \in V$, it is easy to see that each optimal flow can be decomposed into flows on simple cycles. Hence, we can restrict our considerations to flows that are contained in the cone C that is spanned by flows on simple cycles with unit flow value. The result of Theorem 3.5 cannot be applied to this problem for two reasons: On the one hand, since we are dealing with arbitrary costs, it clearly does no longer hold that $c^T x^{(l)}$ is constant among all flows on cycles with unit flow value. On the other hand, any minimizing oracle would be required to return a vector $x^{(l)}$ that minimizes $d^T x^{(l)}$ for a given cost vector d, so it would be necessary to find a most negative cycle C^* in the underlying graph. However, this problem is known to be $\mathcal{NP}$-complete since finding a most negative simple cycle in a graph with edge costs $d_e = -1$ for each $e \in E$ is equivalent to deciding if the graph contains a Hamiltonian cycle (cf. Garey and Johnson (1979)). Nevertheless, we are able to determine a cycle C with the same *sign* as the most negative cycle C^* efficiently by computing a minimum mean cycle in $\mathcal{O}(nm)$ time (cf. (Karp, 1978)). Hence, we can apply Theorem 3.10 to the budget-constrained minimum cost flow problem in order to obtain an FPTAS running in $\mathcal{O}\left(\frac{1}{\varepsilon^2} \cdot m \log m \cdot \left(m + (nm)^2\right)\right)$ time. We will study this problem in detail in the upcoming Chapters 4 and 5 and will improve upon this FPTAS in Section 4.5. ◁

3.3.3 Separation Oracles

We conclude this section by a third kind of oracle called *separation oracle*. Such an oracle embodies the most natural, but also the weakest of the three considered oracles.

Nevertheless, as it will be shown in the following, we can still obtain an FPTAS that runs within the same time bound as the FPTAS given in the previous subsection.

Definition 3.13 (Separation Oracle):
For a given vector $d \in \mathbb{R}^n$, a *separation oracle* for the set S either states that $d^T x^{(i)} \geqslant 0$ for all vectors $x^{(i)} \in S$ or returns a *certificate* $x^{(l)} \in S$ that fulfills $d^T x^{(l)} < 0$. ◁

The name "separation oracle" is based on the fact that such an oracle can in fact be seen as a traditional separation oracle for the *dual cone* $C^* := \{w \in \mathbb{R}^n : w^T x \geqslant 0$ for all $x \in C\}$ of the cone C (cf. (Grötschel et al., 1993)).

Clearly, in comparison to the two oracles introduced before, a separation oracle is the weakest kind of oracle. In particular, the case that $d^T x^{(i)} \geqslant 0$ for all vectors $x^{(i)} \in S$ does no longer include the information whether there is a vector $x^{(l)} \in S$ with $d^T x^{(l)} = 0$ (in which case we have found the desired vector in the parametric search as described above) or if $d^T x^{(i)} > 0$ for all $x^{(i)} \in S$. For example, if we come across a comparison of the form $a_0 + \lambda \cdot a_1 \leqslant b_0 + \lambda \cdot b_1$ during the simulation where $a_1 > b_1$, we are actually interested in the information whether or not the optimal value λ^* fulfills $\lambda^* \leqslant \lambda' := \frac{b_0 - a_0}{a_1 - b_1}$. However, if we use the separation oracle with the cost vector $d(\lambda')$, we only obtain the information whether $\lambda^* < \lambda'$ (in case that the oracle returns a certificate) or if $\lambda^* \geqslant \lambda'$. Hence, in the latter case, the outcome of the comparison is not yet resolved since we need the additional information whether or not $\lambda^* = \lambda'$, so we cannot continue the simulation without any further ado. Nevertheless, as it will be shown in the following theorem, we can gather this additional information by a more sophisticated approach:

Theorem 3.14:
Given a strongly combinatorial and strongly polynomial-time separation oracle $\mathcal{A}$ for S, there is an FPTAS for the problem (3.3) running in $\mathcal{O}\left(\frac{1}{\varepsilon^2} \cdot m \log m \cdot (N + T_{\mathcal{A}}^2)\right)$ time.

Proof: The claim directly follows from Theorem 3.10 if we can show that we can extend the given separation oracle into a sign oracle for S. As in the proof of Theorem 3.10, we simulate the execution of the separation oracle using the parametric cost vector $d(\lambda) := a - \lambda c$. Assume that the execution halts at a comparison that needs to be resolved, resulting in a candidate value λ' for the desired value λ^*. We call the separation oracle with the cost vector $d := d(\lambda')$. Clearly, if the oracle returns a certificate $x^{(l)}$ with $d^T x^{(l)} < 0$, we can conclude that $D(\lambda') < 0$ such that the value λ' was too large according to Lemma 3.8 and the result of the comparison is determined. Conversely, if the oracle states that $d^T x^{(i)} \geqslant 0$ for all $x^{(i)} \in S$, we can conclude that $D(\lambda') \geqslant 0$. However, we may not yet be able to resolve the comparison since its result may rely on the additional information whether $D(\lambda') = 0$ or $D(\lambda') > 0$ as shown above. Nevertheless, we can extract this information by one additional call to the oracle as it will be shown in the following.

First suppose that $\lambda' = \lambda^*$. In this situation, it holds that $d(\lambda')^\mathsf{T} x^{(i)} \geqslant 0$ for all $x^{(i)} \in S$ and there is at least one vector $x^{(l)} \in S$ that fulfills $d(\lambda')^\mathsf{T} x^{(l)} = 0$. Since all the functions $f^{(i)}(\lambda) := d(\lambda)^\mathsf{T} x^{(i)} = a^\mathsf{T} x^{(i)} - \lambda \cdot c^\mathsf{T} x^{(i)}$ are linear functions of λ with negative slope (in case that $c^\mathsf{T} x^{(i)} > 0$; otherwise, the function has no positive root at all), it holds that several functions $f^{(l)}$ evaluate to zero at λ' while every other function attains its root at a larger value for λ (cf. Figure 3.1a). Hence, for every larger value of λ, the separation oracle changes its outcome and returns a certificate. In particular, for a sufficiently small but positive value of δ, the separation oracle called with the cost vector $d(\lambda' + \delta)$ returns a vector $x^{(l)} \in S$ with $d(\lambda' + \delta) x^{(l)} < 0$ that additionally fulfills $d(\lambda')^\mathsf{T} x^{(l)} = 0$ (so $x^{(l)}$ yields the most violated constraint in the overall procedure). Clearly, the value of δ must be small enough to guarantee that we do not reach the root of another function $f^{(i)}$ (i.e., smaller than the distance between the dashed and the dotted line in Figure 3.1a).

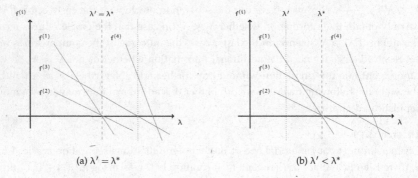

(a) $\lambda' = \lambda^*$ (b) $\lambda' < \lambda^*$

Figure 3.1: Illustration of the two cases that may occur during the simulation of the separation oracle in case that the separation oracle did not return a certificate when evaluated for a candidate value λ'.

Now suppose that $\lambda' < \lambda^*$ (cf. Figure 3.1b). In this case, for a sufficiently small but positive value of δ, the separation oracle returns the *same* answer when called with the cost vector $d(\lambda' + \delta)$ as long as $\lambda' + \delta \leqslant \lambda^*$ (i.e., as long as δ is smaller than the distance between the dotted and the dashed line in Figure 3.1b). Consequently, in order to separate this case from the former case, it suffices to specify a value for δ that is smaller than the distance between any two roots of the functions that occur both in the instance and during the simulation of $\mathcal{A}$. We can then use a second call to the decision oracle in order to decide whether a candidate value λ' is smaller than or equal to the optimal value λ^*.

First note that the root of each function $f^{(i)}$ is given by the rational number $\frac{a^T x^{(i)}}{c^T x^{(i)}}$. Since the coefficients c_j are part of the instance I to GPF and since the values $a_j = \sum_{i=1}^m y_i \cdot A_{ij}$ are generated within the framework of Garg and Koenemann (2007), the encoding length of both values is polynomial in the problem size. Similarly, as noted at the beginning of this section, we can assume that the encoding lengths of the vectors $x^{(i)}$ returned by the oracle are polynomially bounded in n. Consequently, there is some bound M_f with polynomial encoding length such that the root of each function $f^{(i)}$ can be represented by a fraction $\frac{p_i}{q_i}$ with $p_i, q_i \in \mathbb{N}$ and $q_i \leqslant M_f$.

Now consider the root $-\frac{a_0 - b_0}{a_1 - b_1}$ of some function g of the form $g(\lambda) := (a_0 - b_0) + \lambda \cdot (a_1 - b_1)$ that stems from a comparison of two linear parametric values of the forms $a_0 + \lambda \cdot a_1$ and $b_0 + \lambda \cdot b_1$. Assume that we are in the k-th step of the simulation. Since the oracle algorithm is strongly combinatorial, the values $a_0 + \lambda \cdot a_1$ and $b_0 + \lambda \cdot b_1$ result from one or more of the input values $d_j := a_j - \lambda \cdot c_j$ (which are the only linear parametric values at the beginning of the simulation) as well as a sequence of up to k additions or subtractions with other linear parametric values and multiplications with constants. Hence, since $k \in \mathcal{O}(T_A)$ and A runs in (strongly) polynomial time, there is a bound M_g with polynomial encoding length such that the root $-\frac{a_0 - b_0}{a_1 - b_1}$ of each such function g considered up to the k-th step of the simulation can be represented by a fraction of the form $\frac{p}{q}$ with $p, q \in \mathbb{N}$ and $q \leqslant M_g$.

Now let $\mu_1 = \frac{p_1}{q_1}$ and $\mu_2 = \frac{p_2}{q_2}$ with $\mu_1 \neq \mu_2$ denote the roots of two of the above functions of the form $f^{(i)}$ or g. Since $q_1, q_2 \leqslant M_f \cdot M_g$, we get that

$$|\mu_1 - \mu_2| = \left| \frac{p_1}{q_1} - \frac{p_2}{q_2} \right| = \left| \frac{p_1 \cdot q_2 - p_2 \cdot q_1}{q_1 \cdot q_2} \right| \geqslant \frac{1}{M_f^2 \cdot M_g^2} =: \mu$$

Hence, choosing $\delta := \frac{\mu}{2}$, we are able to differentiate between the three cases $D(\lambda) < 0$, $D(\lambda) = 0$, and $D(\lambda) > 0$. Moreover, by returning any[3] vector in S in the case of $D(\lambda) > 0$ and returning the certificate in every other case, the separation oracle is extended into a sign oracle and the correctness follows from the proof of Theorem 3.10. Note that the running-time of the overall procedure remains to be $\mathcal{O}\left(\frac{1}{\epsilon^2} \cdot m \log m \cdot (N + T_A^2) \right)$ (as in the case of a sign oracle in Theorem 3.10) since the encoding length of the number δ is polynomially bounded and since the oracle algorithm is assumed to run in strongly polynomial time. □

---

3   Actually, since we do not have direct access to the set S, we need to obtain such a vector via an oracle access. However, by calling the oracle once more with a very large value for λ or by returning *some* vector found before, we obtain a certificate in S, which we can return.

Example 3.15:

The *budget-constrained minimum cost generalized flow problem* is a generalization of the maximum generalized flow problem that was considered in Section 2.4 in which each edge $e \in E$ has integral costs c_e and in which a budget-constraint $\sum_{e \in E} b_e \cdot x_e \leqslant B$ is added as above. We consider the circulation based version in which $\text{excess}_x(v) = 0$ for each node $v \in V$. As it was shown in (Wayne, 2002), every such generalized circulation x can be decomposed into at most m flows on *unit-gain cycles* and *bicycles*, i.e., flows on cycles C with $\prod_{e \in C} \gamma_e = 1$ and flows on pairs of cycles (C_1, C_2) with $\prod_{e \in C_1} \gamma_e < 1$ and $\prod_{e \in C_2} \gamma_e > 1$ that are connected by a path, respectively. Hence, every generalized circulation lies in the cone that is generated by flows on such unit-gain cycles and bicycles. Wayne (2002) presents a procedure that, for a given cost vector d, either returns a negative cost unit-gain cycle or negative cost bicycle or states that no such cycle exists in $\widetilde{O}(n^2 m)$ time[4]. Thus, by interpreting this algorithm as a separation oracle, we obtain an FPTAS for the budget-constrained minimum generalized flow problem running in $\widetilde{O}\left(\frac{1}{\varepsilon^2} \cdot n^4 m^3\right)$ time according to Theorem 3.14. To the best of our knowledge, this is the first known FPTAS for this problem. $\lhd$

Note that we can also directly extend the separation oracle in Example 3.15 into a sign oracle and apply Theorem 3.10 in this special example: For the case that the algorithm states that there are no negative cost unit-gain cycles and bicycles, Wayne (2002) shows that there are node potentials π such that the reduced costs $d_e^\pi := d_e + \pi_v - \gamma_e \cdot \pi_w$ are non-negative for each edge $e = (v, w) \in E$ with gain factor γ_e. Since $\sum_{e \in E} d_e^\pi \cdot x_e = \sum_{e \in E} d_e \cdot x_e$ for each feasible generalized circulation x (cf. (Wayne, 2002)), there is a unit-gain cycle or bicycle with zero cost if and only if there is a unit-gain cycle or bicycle in the subgraph that is induced by the edges with zero reduced costs. Hence, since unit-gain cycles and bicycles can be detected in $O(nm)$ time according to Wayne (2002), we obtain a sign oracle for the set S of flows on unit-gain cycles and bicycles running in $\widetilde{O}(n^2 m)$ time, which yields an FPTAS with the same running time as above according to Theorem 3.10.

3.4 Conclusion

As it was shown in this chapter, the parametric search technique due to Megiddo (1979, 1983) and the fractional packing framework of Garg and Koenemann (2007) are two generic likewise powerful tools that can be used in order to derive fast algorithms for parametric problems on the one side and to obtain fully polynomial-time

---

4   The notation $\widetilde{O}(\cdot)$ ignores poly-logarithmic factors in m, so $\widetilde{O}(f(n, m)) = O(f(n, m) \cdot \log^k m)$ for some constant k.

approximation schemes for packing problems on the other side. A combination of both frameworks yields an even more powerful framework that can be used in order to solve general packing problems over polyhedral cones. Instead of requiring the cone to be given explicitly, we only need oracle access to the cone, which allows us to extend our results to cones that are generated by an exponential number of vectors. As it was shown, the efficiency of the procedure depends on the power of the corresponding oracle. Moreover, it was shown (and will be shown in the following chapters) that the framework can be applied to a large class of network flow problems. The results of this chapter are summarized in Table 3.1.

Minimizing Oracle	Sign Oracle	Separation Oracle
Theorem 3.5: FPTAS in $\mathcal{O}(\frac{1}{\varepsilon^2} \cdot m \log m \cdot (N + T_A))$ time if $c^T x^{(l)} = \bar{c}$ for each $x^{(l)} \in S$		
Corollary 3.11: FPTAS in $\mathcal{O}(\frac{1}{\varepsilon^2} \cdot m \log m \cdot (N + T_A^2))$ time for strongly combinatorial and strongly polynomial-time $\mathcal{A}$	**Theorem 3.10:** FPTAS in $\mathcal{O}(\frac{1}{\varepsilon^2} \cdot m \log m \cdot (N + T_A^2))$ time for strongly combinatorial and strongly polynomial-time $\mathcal{A}$	**Theorem 3.14:** FPTAS in $\mathcal{O}(\frac{1}{\varepsilon^2} \cdot m \log m \cdot (N + T_A^2))$ time for strongly combinatorial and strongly polynomial-time $\mathcal{A}$

Table 3.1: The summarized results for the generalized packing framework in Chapter 3.

4 Budget–Constrained Minimum Cost Flows: The Continuous Case

In this chapter, we study an extension of the well-known minimum cost flow problem in which a second kind of costs (called *usage fees*) is associated with each edge. The goal is to minimize the first kind of costs as in traditional minimum cost flows while the total usage fee of a flow must additionally fulfill a budget constraint. In the first part of this chapter, we present a specialized network simplex algorithm for the problem. In particular, we provide optimality criteria as well as measurements to avoid cycling. Moreover, we prove a pseudo-polynomial running time of the algorithm using Dantzig's pivoting rule (cf. (Ahuja et al., 1988)). In the second part of the chapter, we present an interpretation of the problem as a bicriteria minimum cost flow problem, which allows us to obtain a weakly polynomial-time combinatorial algorithm that computes only $\mathcal{O}(\log M)$ traditional minimum cost flows, where M is the largest number that occurs in the problem instance. Moreover, we present a strongly polynomial-time algorithm that computes $\widetilde{\mathcal{O}}(nm)$ traditional minimum cost flows and derive three fully polynomial-time approximation schemes for the problem on general and on acyclic graphs.

This chapter is based on joint work with Sven O. Krumke and Clemens Thielen (Holzhauser et al., 2016a, 2015a, 2016b).

4.1 Introduction

As it was motivated in Section 2.4, traditional minimum cost flows provide a useful tool in finding the cheapest way to transport a specific amount of flow through a given network without violating any capacity constraint. Many real-world scenarios, however, require the incorporation of a second kind of costs connected with the usage of the edges that is constrained by a given budget. We study an extension of the traditional minimum cost flow problem by such a second kind of linear costs, called *usage fees*. We show that we obtain both specialized algorithms and efficient approximation schemes for this problem.

The applications of the presented extension of the traditional minimum cost flow problem are manifold. The budget-constrained minimum cost flow problem can be seen as the ε-constraint method applied to a bicriteria minimum cost flow problem (cf., e.g., (Chankong and Haimes, 2008)). As the minimum cost flow problem contains many important network flow problems as a special case such as the maximum flow problem, the dynamic maximum flow problem, the transportation problem, and the assignment problem (cf. (Ahuja et al., 1993)), our algorithms can be directly applied to the case of the budget-constrained variants of these problems. This in turn yields several applications in logistics, telecommunication, and computer networks (cf. (Çalışkan, 2009)). Moreover, our problem is strongly related to the case of the minimum cost flow problem with a fractional objective function of the form $\min \frac{c^T x + c_0}{b^T x + b_0}$ where $b^T x + b_0 > 0$ for each feasible flow x: As it is well known, problems of this form can be solved by applying the so-called *Charnes-Cooper transformation*, which basically replaces the objective function by a linear function of the form $c^T x + c_0 t$ for a new variable t and adds a budget-constraint of the form $b^T x + b_0 t = 1$ (cf. (Charnes and Cooper, 1962)).

In the subsequent chapter, we will investigate two discrete variants of the budget-constrained minimum cost flow problem. Both the continuous version of the problem considered in this chapter and the two discrete variants can be seen as a *network improvement problem* if we interpret the usage fees as the costs that are necessary to *upgrade* the capacity of the corresponding edges up to a sufficiently large amount. As we will show in the next chapter, this interpretation allows to model even more realistic applications.

4.1.1 Previous Work

The related budget-constrained maximum flow problem was first studied by Ahuja and Orlin (1995), who present a weakly polynomial-time algorithm for the problem that is based on a capacity scaling variant of the successive shortest path algorithm. Çalışkan (2009) later showed that Ahuja and Orlin's (1995) algorithm may not return a feasible solution in a specific special case and presented a corrected version of the algorithm. The same author also presented a double scaling algorithm, a network simplex algorithm, and a cost scaling algorithm for the budget-constrained maximum flow problem and evaluated their empirical performance (cf. (Çalışkan, 2008, 2011, 2012)). In particular, he could show that the cost scaling variant outperforms the other two scaling variants (including the capacity scaling algorithm of Ahuja and Orlin (1995)) and that the network simplex algorithm clearly outperforms all known algorithms for the problem (including CPLEX applied to the LP formulation). Krumke and Schwarz (1998) study the problem of finding a maximum flow in the case that the capacity of each edge can be improved using a given budget. To do so, they differentiate between three variants of how to calculate the upgrade costs. One of these variants is equivalent to the one that is used in this chapter. The other two variants will be investigated in the subsequent chapter.

In (Demgensky et al., 2002) and (Demgensky et al., 2004), the authors deal with a related problem in which the costs and/or the capacity of each edge can be improved by investing money. The aim is to maximize the flow value while not exceeding the given budget and a bound on the costs of the flow. They show that the problems are $\mathcal{NP}$-complete and provide pseudo-polynomial-time algorithms for the problems on series-parallel graphs. Similar research was previously done by Krumke et al. (1998), who provide several results about the complexity and approximability of models in which either edges or nodes can be improved in order to reduce the latency of the corresponding (incident) edges.

4.1.2 Chapter Outline

We investigate the budget-constrained minimum cost flow problem with respect to efficient combinatorial solution methods and fast approximation schemes. After a provision with the necessary preliminaries in Section 4.2, we concentrate on a specialized network simplex algorithm for the budget-constrained minimum cost flow problem in Section 4.3. In particular, we show how the definition of a basis structure and how the simplex pivot step itself need to be adopted in comparison to the network simplex algorithm for the traditional minimum cost flow problem (cf. (Ahuja et al., 1993)). Moreover, we present optimality criteria for the problem that are based on a novel definition of the reduced costs of each edge. We combine two tools that are used in the traditional network simplex algorithm to avoid cycling of the procedure in order to obtain an equivalent result for the presented algorithm. In particular, we show that the number of degenerate simplex pivots between two non-degenerate pivots is pseudo-polynomially bounded, which shows that the procedure terminates within pseudo-polynomial time. In the second part of this chapter (Section 4.4), we present an interpretation of the budget-constrained minimum cost flow problem as a bicriteria minimum cost flow problem and show that we can reduce the problem to the determination of a suitable weighting of the costs and the usage fees. This in turn allows us to obtain a weakly polynomial-time algorithm for the problem running in $\mathcal{O}(\log M \cdot \text{MCF}(m, n, C, U))$ time, where M is the largest number in the instance and MCF is defined as in Section 2.4. Incorporating the same observation into Megiddo's parametric search technique as introduced in Section 3.1, we obtain a strongly polynomial-time algorithm that computes at most $\widetilde{\mathcal{O}}(nm)$ traditional minimum cost flows. Finally, in Section 4.5, we consider the approximability of the budget-constrained minimum cost flow problem and present two FPTASs for the problem on general graphs. Moreover, we specialize one of these FPTASs to the case of acyclic graphs, which allows us to obtain a more efficient running time. An overview of the results of this chapter is given in Table 4.1 on page 76.

4.2 Preliminaries

We start by defining the constrained minimum cost flow problem in a directed graph $G = (V, E)$ with *edge capacities* $u_e \in \mathbb{N}$, *edge costs* $c_e \in \mathbb{Z}$ (i.e., we allow integral costs with arbitrary sign), and *usage fees* $b_e \in \mathbb{N}_{\geq 0}$ per unit of flow on the edges $e \in E$, as well as a *budget* $B \in \mathbb{N}_{\geq 0}$ and a distinguished *source* $s \in V$ and *sink* $t \in V$.[1]

Definition 4.1 (Flow, flow value, constrained minimum cost flow):
A function $x \colon E \to \mathbb{R}_{\geq 0}$ is called a (feasible) *flow* if $x_e := x(e) \leq u_e$ for each $e \in E$, $\mathrm{excess}_x(v) := \sum_{e \in \delta^-(v)} x_e - \sum_{e \in \delta^+(v)} x_e = 0$ for each $v \in V \setminus \{s, t\}$ and $b(x) := \sum_{e \in E} b_e \cdot x_e \leq B$. The *flow value* of a flow x is given by $\mathrm{val}(x) := \mathrm{excess}_x(t)$. A flow x of minimum cost $c(x) := \sum_{e \in E} c_e \cdot x_e$ is called a *budget-constrained minimum cost flow* or just *optimal flow*. ◁

Note that we explicitly allow the existence of parallel edges between two nodes since these edges may differ in their capacities, costs, and usage fees, which may be useful in certain applications, e.g., in order to model piecewise linear cost functions (cf. (Krumke and Schwarz, 1998)). In the presented *continuous case* of the budget-constrained minimum cost flow problem, the total usage fee on an edge e that transports x_e units of flow is given by $b_e \cdot x_e$, so the overall usage fee $b(x)$ of a flow x is given by $b_\mathbb{R}(x) := \sum_{e \in E} b_e \cdot x_e$. This is probably the most natural formulation of the problem and corresponds to the constraint used by Ahuja and Orlin (1995) and Çalışkan (2008, 2011, 2012).

Definition 4.2 (Continuous Budget-constrained minimum cost flow problem (BCMCFP$_\mathbb{R}$)):

INSTANCE: Directed graph $G = (V, E)$ with source $s \in V$, sink $t \in V$, capacities $u_e \in \mathbb{N}_{\geq 0}$, costs $c_e \in \mathbb{Z}$, and usage fees $b_e \in \mathbb{N}_{\geq 0}$ on the edges $e \in E$ and a budget $B \in \mathbb{N}_{\geq 0}$.

TASK: Determine a budget-constrained minimum cost flow for $b := b_\mathbb{R}$. ◁

---

1 As it is common when dealing with network flow problems, we assume throughout this chapter that all of these values are integral, which is no restriction for most of the applications since we can multiply all values with their least common denominator in case of rational data (cf., e.g., (Ahuja et al., 1993)).

Throughout this chapter, we make the following assumption on the structure of the underlying graph:

Assumption 4.3: Every node $v \in V$ is contained in some directed path from the source s to the sink t. ◁

Note that Assumption 4.3 does not impose any restriction on the underlying model since we can connect the source and the sink with every node $v \in V$ by an edge with large costs, capacity, and usage fees if necessary. As a consequence, we can assume the underlying graph to be weakly connected, so $n \in \mathcal{O}(m)$.

In the following, when making statements about *time complexities*, we pinpoint that we use the definitions $C := \max_{e \in E} |c_e|$, $U := \max_{e \in E} u_e$, and $B := \max_{e \in E} b_e$ in accordance to common notation (cf., e.g., (Ahuja et al., 1988; Ahuja and Orlin, 1995; Çalışkan, 2008)). Moreover, we define $M := \max\{m, C, U, B\}$ such that the encoding length of any instance of BCMCFP$_\mathbb{R}$ is contained in $\mathcal{O}(m \log M)$ (cf., e.g., (Ahuja and Orlin, 1995)). In all other cases, we usually refer to cycles by C and denote the budget itself by B.

Using the definitions from above, we can formulate the budget-constrained minimum cost flow problem as a linear program as follows:

$$\min \sum_{e \in E} c_e \cdot x_e \tag{4.1a}$$

$$\text{s.t.} \sum_{e \in \delta^-(v)} x_e - \sum_{e \in \delta^+(v)} x_e = 0 \qquad \text{for all } v \in V \setminus \{s, t\}, \tag{4.1b}$$

$$\sum_{e \in E} x_e \cdot b_e \leqslant B, \tag{4.1c}$$

$$0 \leqslant x_e \leqslant u_e \qquad \text{for all } e \in E. \tag{4.1d}$$

Clearly, since the mathematical model (4.1a) – (4.1c) is a linear program, it can be solved in weakly polynomial time by known techniques such as interior point methods (cf. (Schrijver, 1998)). In particular, Vaidya (1989) shows that the traditional minimum cost flow problem can be solved in $\mathcal{O}(n^2 \cdot m^{0.5} \cdot \log M')$ time if M' is the largest number that occurs in the corresponding instance. This result relies on the fact that the problem can be formulated as a linear program of the form $\max\{w^\mathsf{T} x : Gx = 0, Hx \geqslant b\}$ for a block matrix H and a matrix G with at most a constant number of non-zero entries per column. Any instance of BCMCFP$_\mathbb{R}$ fulfills this property as well after inserting a slack variable and incorporating the budget constraint into the matrix G. For the case of simple graphs, this leads to a running-time of $\mathcal{O}(n^2 \cdot m^{0.5} \cdot \log M)$. Since we can transform each simple graph into a multigraph by placing an artificial node in the middle of each edge as described in Section 2.4, we get the following weakly polynomial running time for BCMCFP$_\mathbb{R}$ on the more general setting of multigraphs:

Theorem 4.4:
BCMCFP$_\mathbb{R}$ is solvable in weakly polynomial time $\mathcal{O}(m^{2.5} \cdot \log M)$. $\qquad$ $\square$

However, in this thesis, we are interested in *combinatorial* algorithms that exploit the discrete structure of the underlying problem. We show how we can obtain tailored algorithms and how we can incorporate combinatorial algorithms for the traditional minimum cost flow problem in order to solve the more general budget-constrained minimum cost flow problem BCMCFP$_\mathbb{R}$.

Note that $c(x^*) \leqslant 0$ for each optimal flow x^* since we do not incorporate nontrivial lower bounds on the flow on the edges or require some specific amount of flow to pass the network, so the zero flow is always feasible. Nevertheless, as it was shown in Section 2.4, we are able to model a required flow value F if we append an artificial edge with capacity F and sufficiently small costs to the former sink t that leads to a new sink t'. The fact that $c(x^*) \leqslant 0$ for each optimal flow x^* has the consequence that we need to adopt the definition of an FPTAS according to Section 2.5 appropriately. That means, we need to handle BCMCFP$_\mathbb{R}$ as a maximization problem rather than a minimization problem when speaking about approximation algorithm, so a solution x returned by an FPTAS with respect to a precision parameter $\varepsilon \in (0, 1)$ fulfills $(-c(x)) \geqslant (1 - \varepsilon) \cdot (-c(x^*))$ or, equivalently, $c(x) \leqslant (1 - \varepsilon) \cdot c(x^*)$.

Consider a traditional minimum cost flow with respect to the costs c that is computed by some state-of-the-art algorithm, for example the enhanced capacity scaling algorithm by Orlin (1993) running in $\mathcal{O}(m \log m \cdot (m + n \log n))$ time on multigraphs as it will be shown in Section 4.4. If the total usage fee of this flow fulfills $b(x) \leqslant B$, we have clearly found an optimal solution to the given instance of the budget-constrained minimum cost flow problem and are done. In the following, we are interested in the converse case that the budget is exceeded for this flow. Note that the usage fee amounts to at least $B + 1$ in this case since the computed traditional minimum cost flow can be assumed to be integral without loss of generality and since the usage fees are integral as well:

Assumption 4.5: There is a traditional minimum cost flow x with respect to the costs c that fulfills $\sum_{e \in E} b_e \cdot x_e \geqslant B + 1$. $\qquad$ $\lhd$

4.3 Network Simplex Algorithm

Since it was published by Dantzig (1951) (originally designed for the transportation problem), the network simplex algorithm for the traditional minimum cost flow problem has been improved progressively and is widely believed to be one of the most efficient solution methods for the minimum cost flow problem at present (cf. (Ahuja et al., 1993)). Simplex-type methods usually have to cope with the risk to "get stuck" in an infinite loop with no progress – an effect that is referred to as *cycling*. However, Cunningham (1976) introduced the notion of *strongly feasible bases* that may be used to prevent cycling. While the sequence of operations with no progress may still be exponentially large, several authors such as Cunningham (1979) and Ahuja et al. (2002) provide measurements to keep the length of such sequences polynomially bounded. At present, the network simplex algorithm with the best time complexity is due to Orlin (1997) in combination with Tarjan's (1997) dynamic tree data structure and achieves a running time of $O(nm \log n \min\{\log nC, m \log n\})$.

In this section, we show how we can adopt several of the above ideas in order to obtain a network simplex algorithm for the budget-constrained minimum cost flow problem running in pseudo-polynomial time. In particular, we show how we can adapt the mentioned measurements to avoid cycling to the more complex basis structures that we need to cope with in the case of $BCMCFP_\mathbb{R}$. Independently, Çalışkan (2011) published a network simplex algorithm for the strongly related budget-constrained maximum flow problem. He could show that his implementation clearly outperforms other solution methods including CPLEX by a factor of 27 on average in terms of running times. Since the constraints of the (budget-constrained) maximum and minimum-cost flow problem are the same, his algorithm can be applied to the case of $BCMCFP_\mathbb{R}$ with only minor modifications. Nevertheless, the upcoming results complement the ones of Çalışkan in several aspects: On the one hand, we provide a fully combinatorial description of the algorithm that does not rely on any linear programming arguments. On the other hand, in contrast to the algorithm shown in (Çalışkan, 2011), our algorithm is based on *two* different kinds of (integral) node potentials and *three* kinds of reduced costs. We provide a proof that these reduced costs may in fact be used as optimality criteria for the procedure. Moreover, the integral node potentials enable us to show that our procedure can be implemented to run in pseudo-polynomial time. As it will be moreover shown, this running time can be further improved by incorporating Dantzig's pivoting rule for choosing the edge that enters the basis (cf., e.g., (Ahuja et al., 1988)). Finally, in his measurements to avoid cycling, Çalışkan (2011) misses to include all cases that may occur. We close this gap by introducing a novel transformation step that allows us to avoid the corresponding case and to simplify the proofs.

4.3.1   Notation and Definitions

In the following, we give insights into the notion of *basis structures* in the context of BCMCFP$_\mathbb{R}$. In contrast to the network simplex algorithm for the traditional minimum cost flow problem (cf. (Ahuja et al., 1993)), we need to drop the assumption that the subgraph that is induced by edges that are neither empty nor full is cycle free. Instead, the basis contains a cycle with non-zero usage fees, as it will be shown in detail in the following.

Consider an edge $\bar{e} \in E$ and a partition of the remaining edges in $E \setminus \{\bar{e}\}$ into three sets L, T, and U. Let the edges in T form a spanning-tree of the underlying graph G. Since there is a unique (undirected) path between any two nodes in the subgraph induced by T, each edge $e \in L \cup U \cup \{\bar{e}\}$ closes a unique (undirected) cycle C(e) together with the edges in T. In the following, for each $e \in L \cup U \cup \{\bar{e}\}$, let $C^+(e)$ ($C^-(e)$) denote the set of edges that are oriented in the same (opposite) direction as e in C(e). The costs and usage fees of this cycle are then given by $c(C(e)) := \sum_{e \in C^+(e)} c_e - \sum_{e \in C^-(e)} c_e$ and $b(C(e)) := \sum_{e \in C^+(e)} b_e - \sum_{e \in C^-(e)} b_e$, respectively[2].

In particular, note that the subgraph induced by the edges in $T \cup \{\bar{e}\}$ contains a cycle $C(\bar{e})$. We call such a tuple $(L, T, U, \bar{e})$ a *basis structure* of the budget-constrained minimum cost flow problem if $b(C(\bar{e})) \neq 0$.

For a given basis structure $(L, T, U, \bar{e})$, let x denote a flow that fulfills $x_e = 0$ for each $e \in L$, $x_e = u_e$ for each $e \in U$, and $b(x) := \sum_{e \in E} b_e \cdot x_e = B$ while maintaining flow conservation at each node $v \in V$. We refer to x as the *basic solution* corresponding to $(L, T, U, \bar{e})$. As shown in (Çalışkan, 2011), the basic solution of each basis structure is uniquely defined and can be obtained in $\mathcal{O}(m)$ time: In a first step, we let $x_{\bar{e}} = 0$ and determine the values of all edges in T as it is done in the traditional network simplex method (cf. (Ahuja et al., 1993)). In a second step, the flow on the cycle $C(\bar{e})$ is then increased (or decreased) until $b(x) = B$. In the following, we refer to a basis structure $(L, T, U, \bar{e})$ as *feasible* if the corresponding basic solution x is a feasible flow. In this case, we also refer to the flow x as a *basic feasible flow*. The described procedure yields the following corollary:

Corollary 4.6:
For each feasible basis structure $(L, T, U, \bar{e})$ and its corresponding basic feasible flow x,

---

2   As noted above, the meaning of the variables C, U and B is ambiguous. However, in order to comply with the common notations (cf., e.g., (Ahuja et al., 1988; Ahuja and Orlin, 1992)), we pinpoint that we use the definitions $C := \max_{e \in E} |c_e|$, $U := \max_{e \in E} u_e$, and $B := \max_{e \in E} b_e$ when making statements about *time complexities*, but refer to cycles, the set of full edges, and the budget, respectively, in all other cases.

it holds that x can be decomposed into two flows x^I and x^C such that $x = x^I + x^C$, where x^I is integral and x^C is only positive on $C(\bar{e})$. $\qquad\square$

Note that Corollary 4.6 holds independently of whether the budget B is integral or not (this will be important in Section 4.3.4).

As it turns out, we are able to restrict our considerations to such feasible basis structures and their corresponding basic feasible flows. This is shown in the following lemma, whose proof we include for the sake of completeness:

Theorem 4.7 (Ahuja et al. (1993)):
Each optimal solution x of $\text{BCMCFP}_{\mathbb{R}}$ can be turned into an optimal basic feasible flow x^* in $\mathcal{O}(m^2)$ time.

Proof: Let x be an optimal solution to the underlying instance of $\text{BCMCFP}_{\mathbb{R}}$. Furthermore, let G' denote the subgraph that is induced by the set of *free edges* e fulfilling $0 < x_e < u_e$. Clearly, if G' does not contain every node $v \in V$ or does not contain a cycle, we can add further edges and the claim follows. Else, if there are at least two cycles C_1 and C_2 consisting of free edges, we fix arbitrary orientations of the two cycles. If one of these cycles fulfills $b(C_i) = 0$ for $i \in \{1, 2\}$, it clearly holds that $c(C_i) = 0$ as well (since we could otherwise improve the flow x by sending flow on C_i in some direction without influencing the budget) and we can increase or decrease the flow on C_i until at least one edge becomes empty or full. Similarly, if $b(C_1) \neq 0 \neq b(C_2)$, by sending a sufficiently small positive or negative amount of flow δ on C_1 and reducing or increasing the flow on C_2 by $\delta \cdot \frac{b(C_1)}{b(C_2)}$, respectively, we maintain feasibility (since all of the edges are free edges and the usage fee of the flow remains equal to B) and optimality (since $c(C_1) + \delta \cdot \frac{c(C_1)}{c(C_2)} = 0$ as in the previous case). As before, by sending the maximum possible amount δ such that no flow bound is violated at any edge, the flow on at least one edge e of the edges in C_1 or C_2 becomes zero or equal to u_e, so this edge e can be assigned to L or U, respectively. In any case, the number of free edges in G' decreases by at least one. By repeating the above procedure, we end with an optimal basic feasible flow within $\mathcal{O}(m)$ iterations. Note that we can determine a pair of (not necessarily edge-disjoint) cycles in $\mathcal{O}(m)$ time by a single traversal of the graph and cancel one of these cycles within the same time bound, which yields the claim. $\qquad\square$

As in the traditional network simplex algorithm, we associate *node potentials* with each node $v \in V$ in order to be able to check for optimality quickly. However, since we are dealing with two kinds of costs, we maintain *two* different node potentials π and μ that are defined with respect to the edge costs c_e and the usage fees b_e, respectively. In particular, we define $\pi_s := 0$ and $\mu_s := 0$ for the source s of the network (which we will select as the root node of the spanning tree T in the following). We choose the node potentials π and μ in a way such that the *reduced costs* $c_e^{\pi} := c_e - \pi_v + \pi_w$ and $b_e^{\mu} := b_e - \mu_v + \mu_w$ are zero for each edge $e = (v, w) \in T$. With this restriction, the node potentials at each node $v \in V$ are uniquely defined and can be computed in $\mathcal{O}(n)$ time (cf. (Ahuja et al., 1993) for further details).

Note that, for any edge $e \notin T$, the costs and usage fees of the cycle $C(e)$ are given by

$$
\begin{aligned}
c(C(e)) &= \sum_{e' \in C^+(e)} c_{e'} - \sum_{e' \in C^-(e)} c_{e'} \\
&= \left(\sum_{e'=(v,w) \in C^+(e)} c_{e'} - \pi_v + \pi_w \right) - \left(\sum_{e'=(v,w) \in C^-(e)} c_{e'} - \pi_v + \pi_w \right) \\
&= \sum_{e' \in C^+(e)} c_{e'}^{\pi} - \sum_{e' \in C^-(e)} c_{e'}^{\pi} = c_e^{\pi}
\end{aligned}
$$

and

$$
b(C(e)) = \sum_{e' \in C^+(e)} b_{e'}^{\mu} - \sum_{e' \in C^-(e)} b_{e'}^{\mu} = b_e^{\mu},
$$

respectively, since $c_{e'}^{\pi} = b_{e'}^{\mu} = 0$ for each $e' \in T$ and since $e \in C^+(e)$. Thus, both values $c(C(e))$ and $b(C(e))$ can be computed in constant time once the node potentials are known.

In order to be able to decide if a basic feasible flow is optimal or to detect an edge that is able to improve the objective function, we assign a third kind of reduced costs $d_e^{\pi,\mu}$ to each edge $e \in E$ that is defined as follows:

$$
d_e^{\pi,\mu} := c_e^{\pi} - c_{\bar{e}}^{\pi} \cdot \frac{b_e^{\mu}}{b_{\bar{e}}^{\mu}}. \tag{4.2}
$$

Remember that $b_{\bar{e}}^{\mu} = b(C(\bar{e})) \neq 0$ in any basis structure. Intuitively, the reduced costs $d_e^{\pi,\mu}$ describe the effect that an increase of the flow on $C(e)$ by one unit and a decrease of the flow on $C(\bar{e})$ by $\frac{b_e^{\mu}}{b_{\bar{e}}^{\mu}}$ units has on the objective function value. This will be shown in the following section. Note that $d_e^{\pi,\mu} = 0$ for each $e \in T \cup \{\bar{e}\}$ since $c_e^{\pi} = 0$ and $b_e^{\mu} = 0$ for each $e \in T$ and $d_{\bar{e}}^{\pi,\mu} = c_{\bar{e}}^{\pi} - c_{\bar{e}}^{\pi} \cdot \frac{b_{\bar{e}}^{\mu}}{b_{\bar{e}}^{\mu}} = 0$.

4.3.2 Network Simplex Pivots

Before we describe the network simplex pivot in the case of $\text{BCMCFP}_{\mathbb{R}}$, it is useful to recall the basic outline of the corresponding procedure in the case of the traditional network simplex algorithm: For a given basis structure (L, T, U) that consists of a set of empty edges L, a spanning tree T, and a set of full edges U, assume that there is an edge $e \in L$ with negative reduced costs. Adding this *entering edge* e to the spanning tree T closes a unique cycle $C(e)$ with negative costs. By sending flow on $C(e)$ in the direction of e, we can improve the objective function value until, for some flow value δ, some *leaving edge* $e' \in C(e)$ becomes empty or full. By assigning this edge e' to L or U, respectively, we obtain a new basis structure. This operation (adding an edge to T, sending flow on the cycle, removing one edge from the cycle) is called a *simplex pivot*. One distinguishes between a *degenerate simplex pivot* if $\delta = 0$ and a *non-degenerate simplex pivot* if $\delta > 0$. Note that the objective function value never increases during a simplex pivot, but only decreases in the case of a non-degenerate simplex pivot.

Now, for a given instance of $BCMCFP_R$, let $(L, T, U, \bar{e})$ and x denote a feasible basis structure and its basic feasible flow, respectively, and let π and μ denote the corresponding node potentials. Assume that there is an edge $e \in L$ with negative reduced costs $d_e^{\pi, \mu} < 0$. We show that we do not increase the objective function value if we add the then called *entering edge* e to T (which closes a new cycle $C(e)$ together with the edges in T) and send suitable amounts of flow on *both* of the cycles $C(e)$ and $C(\bar{e})$ until the flow value on at least one *leaving edge* $e' \in T \cup \{e, \bar{e}\}$ becomes equal to zero or $u_{e'}$. In this case, we can obtain a new basis structure $(L', T', U', \bar{e}')$ and continue the procedure.

For some value $\delta \geqslant 0$, let x' be the flow defined as

$$x' := x + \delta \cdot \chi(C(e)) - \delta \cdot \frac{b_e^{\mu}}{b_{\bar{e}}^{\mu}} \cdot \chi(C(\bar{e})), \tag{4.3}$$

where, for any cycle C with forward edges C^+ and backward edges C^-, the flow $\chi(C)$ is defined as

$$(\chi(C))_e := \begin{cases} 1, & \text{if } e \in C^+, \\ -1, & \text{if } e \in C^-, \\ 0, & \text{else.} \end{cases}$$

The new flow x' fulfills $b(x') = B$, since

$$b(x') = b(x) + \delta \cdot b(\chi(C(\dot{e}))) - \delta \cdot \frac{b_e^{\mu}}{b_{\bar{e}}^{\mu}} \cdot b(\chi(C(\bar{e})))$$

$$= b(x) + \delta \cdot b(C(e)) - \delta \cdot \frac{b_e^{\mu}}{b_{\bar{e}}^{\mu}} \cdot b(C(\bar{e}))$$

$$= b(x) + \delta \cdot \left(b_e^{\mu} - \frac{b_e^{\mu}}{b_{\bar{e}}^{\mu}} \cdot b_{\bar{e}}^{\mu} \right) = b(x) = B.$$

Moreover, it holds that

$$c(x') = c(x) + \delta \cdot c(\chi(C(e))) - \delta \cdot \frac{b_e^\mu}{b_{\bar{e}}^\mu} \cdot c(\chi(C(\bar{e})))$$

$$= c(x) + \delta \cdot c(C(e)) - \delta \cdot \frac{b_e^\mu}{b_{\bar{e}}^\mu} \cdot c(C(\bar{e}))$$

$$= c(x) + \delta \cdot \left(c_e^\pi - \frac{b_e^\mu}{b_{\bar{e}}^\mu} \cdot c_{\bar{e}}^\pi \right) = c(x) + \underbrace{\delta}_{\geqslant 0} \cdot \underbrace{d_e^{\pi,\mu}}_{<0} \leqslant c(x).$$

By sending a small amount of $\delta \geqslant 0$ units of flow on $C(e)$ and $-\frac{b_e^\mu}{b_{\bar{e}}} \cdot \delta$ units of flow on $C(\bar{e})$, we do not increase the objective value while maintaining feasibility. In fact, if we can choose a positive value for δ, the objective value strictly decreases. Let $\theta_{e'} := (\chi(C(e)))_{e'} - \frac{b_e^\mu}{b_{\bar{e}}^\mu} \cdot (\chi(C(\bar{e})))_{e'}$ denote the effect that an augmentation of one unit of flow on $C(e)$ and $-\frac{b_e^\mu}{b_{\bar{e}}}$ units of flow on $C(\bar{e})$ has on edge $e' \in E$. Moreover, let δ be defined as $\delta := \min_{e' \in E} \delta_{e'}$ with

$$\delta_{e'} := \begin{cases} -\frac{x_{e'}}{\theta_{e'}} & \text{if } \theta_{e'} < 0, \\ \frac{u_{e'} - x_{e'}}{\theta_{e'}} & \text{if } \theta_{e'} > 0, \\ +\infty & \text{else.} \end{cases}$$

Hence, by sending δ units of flow on $C(e)$ and $-\frac{b_e^\mu}{b_{\bar{e}}}$ units of flow on $C(\bar{e})$, we maintain feasibility of the flow. Moreover, by the definition of δ, there are several *blocking edges* e' contained in $C(e)$ or $C(\bar{e})$ (or both) that fulfill $\delta_{e'} = \delta$. We choose one of these blocking edges as the *leaving edge* e', which consequently fulfills $x'_{e'} = 0$ or $x'_{e'} = u_{e'}$. We distinguish three cases:

- If $e' = e$, we can simply remove e from L and assign it to U. The new basis structure $(L', T', U', \bar{e}') := (L \setminus \{e\}, T, U \cup \{e\}, \bar{e})$ is then feasible again.

- If $e' = \bar{e}$, we obtain a new basis structure by setting $(L', T', U', \bar{e}') := (L \cup \{e'\} \setminus \{e\}, T, U, e)$ or $(L', T', U', \bar{e}') := (L \setminus \{e\}, T, U \cup \{e'\}, e)$, depending on whether $x'_{e'} = 0$ or $x'_{e'} = u_{e'}$, respectively.

- Otherwise, we remove e from L and assign it to T. Moreover, we remove e' from T and assign it to L or U, depending on whether $x'_{e'} = 0$ or $x'_{e'} = u_{e'}$, respectively, which yields the new basis structure $(L', T', U', \bar{e}') := (L \cup \{e'\} \setminus \{e\}, T \cup \{e\} \setminus \{e'\}, U, \bar{e})$ or $(L', T', U', \bar{e}') := (L \setminus \{e\}, T \cup \{e\} \setminus \{e'\}, U \cup \{e'\}, \bar{e})$, respectively. Furthermore, for the case that $\bar{e}$ is no longer contained in a cycle in $T' \cup \{\bar{e}\}$, we assign $\bar{e}$ to T', remove e from T', and set $\bar{e}' := e$.

In any case, we maintain a spanning tree T and ensure that $\bar{e}$ closes a cycle with the edges in T. As in the traditional network simplex algorithm, we refer to such a step as a *simplex pivot*. This pivot step is called *degenerate* if $\delta = 0$ and *non-degenerate* else. In the former case, we refer to those edges with $\delta_e = 0$ as *degenerate edges*. Note that the objective function strictly decreases only in the case of a non-degenerate simplex pivot.

The case that there is an edge $e \in U$ with $d_e^{\pi,\mu} > 0$ is similar to the above case. By decreasing the flow on $C(e)$ by δ units and increasing the flow on $C(\bar{e})$ by $\frac{b_e^\mu}{b_{\bar{e}}^\mu}$ units, i.e., by setting

$$x' := x - \delta \cdot \chi(C(e)) + \delta \cdot \frac{b_e^\mu}{b_{\bar{e}}^\mu} \cdot \chi(C(\bar{e})), \qquad (4.4)$$

we maintain feasibility and improve the objective function for the case that $\delta > 0$.

In the above discussion, we have assumed that $(L, T, U, \bar{e})$ is a (feasible) basis structure, which includes that $b_{\bar{e}}^\mu \neq 0$, i.e., that the usage fee of the cycle $C(\bar{e})$ is non-zero. As it turns out, the usage fee of $C(\bar{e})$ remains non-zero after a simplex pivot as long as it was non-zero before the step, as it is shown in the following lemma:

Lemma 4.8:
Assume that $(L, T, U, \bar{e})$ is a feasible basis structure of the underlying instance of $\text{BCMCFP}_\mathbb{R}$ and let π and μ denote the corresponding node potentials. Then the tuple $(L', T', U', \bar{e}')$ that results from a simplex pivot is a feasible basis structure again.

Proof: Let e denote the entering edge in the simplex pivot that leads to the new basis structure $(L', T', U', \bar{e}')$. As it was shown above, the flow that is induced by this new basis structure is feasible, again. Now assume for the sake of contradiction that $b_{\bar{e}'}^{\mu'} = 0$.

First, consider the case that the two cycles $C(e)$ and $C(\bar{e})$ are edge-disjoint. Clearly, since either one of the edges in $C(e)$ or one of the edges in $C(\bar{e})$ becomes the leaving edge, one of the two cycles remains in $T' \cup \{\bar{e}'\}$. Since $b(C(\bar{e})) = b_{\bar{e}}^\mu \neq 0$ by assumption, it must hold that $b(C(e)) = b_e^\mu = 0$. However, in this case, the flow on the cycle $C(\bar{e})$ does not change according to equations (4.3) and (4.4), i.e., the leaving edge must be contained in $C(e)$ and the cycle $C(\bar{e})$ remains in $T' \cup \{\bar{e}'\}$, which contradicts the assumption that $b_{\bar{e}'}^{\mu'} = 0$.

Now assume that the two cycles $C(e)$ and $C(\bar{e})$ are not edge-disjoint. It is easy to see, that there is exactly one (undirected) simple path P_0 that is contained in both $C(e)$ and $C(\bar{e})$ and that, consequently, contains neither e nor $\bar{e}$ (cf. (Çalışkan, 2011)). The leaving edge e' must then be contained in P_0, which can be seen as follows: For the case that $b_e^\mu \neq 0$, none of the two cycles $C(e)$ and $C(\bar{e})$ can still be contained in $T' \cup \{\bar{e}'\}$ (otherwise, it would again hold that $b_{\bar{e}'}^{\mu'} \neq 0$), i.e., the leaving edge must be a common edge of the two cycles, which lies on P_0. Else, if $b_e^\mu = 0$, the leaving edge e' must be contained in $C(e)$ since the flow does not change on $C(\bar{e})$ as shown above. If e' was not contained in $C(\bar{e})$ as well, the cycle $C(\bar{e})$ would continue to exist in $T' \cup \{\bar{e}'\}$, which, again, contradicts the assumption that $b_{\bar{e}'}^{\mu'} = 0$. Hence, we obtain that $e' \in P_0$. Let v and w denote the end nodes of P_0 and let $P(e)$ and $P(\bar{e})$ denote the unique paths that connect w with v in $C(e) \setminus P_0$ and $C(\bar{e}) \setminus P_0$, respectively. For some fixed orientation of $P(e)$ and $P(\bar{e})$, the new cycle $C(\bar{e}')$ is the concatenation of $P(e)$ and the reversal of $P(\bar{e})$ (cf. Figure 4.1). Since $b_{\bar{e}'}^{\mu'} = 0$ by assumption, it holds that $b(P(e)) = b(P(\bar{e}))$, which implies that

$$b_e^\mu = b(C(e)) = b(P(e)) + b(P_0) = b(P(\bar{e})) + b(P_0) = b(C(\bar{e})) = b_{\bar{e}}^\mu.$$

However, according to equations (4.3) and (4.4), this implies that the flow on $x_{e'}$ remains unchanged, which contradicts the assumption that e' is the leaving edge. □

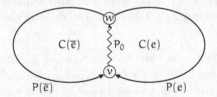

Figure 4.1: The situation if the two cycles $C(e)$ and $C(\bar{e})$ are not edge-disjoint. Since the leaving edge e' lies on the path P_0, the resulting cycle (thick) is the concatenation of $P(\bar{e})$ and $P(e)$.

4.3.3 Optimality Conditions

The above discussion shows that, whenever we encounter an edge $e \in L$ with $d_e^{\pi,\mu} < 0$ or an edge $e \in U$ with $d_e^{\pi,\mu} > 0$, we can perform a simplex pivot and improve the objective function value (in the case that the pivot is non-degenerate). Conversely, as it turns out, whenever there are no such edges in a feasible basis structure, the corresponding feasible basic flow is an optimal solution to the underlying instance of BCMCFP$_\mathbb{R}$. In order to prove this result, we need the following lemma:

Lemma 4.9:
Let (π, μ) denote the node potentials corresponding to the basis structure $(L, T, U, \bar{e})$. Any flow $\bar{x}$ with $b(\bar{x}) = B$ is optimal for the underlying instance of BCMCFP$_\mathbb{R}$ if and only if it is optimal with respect to the objective function $d^{\pi,\mu}(x) := \sum_{e \in E} d_e^{\pi,\mu} \cdot x_e$.

Proof: Using equation (4.2) and the fact that $\sum_{e \in E} c_e^\pi \cdot \bar{x}_e = \sum_{e \in E} c_e \cdot \bar{x}_e$ and $\sum_{e \in E} b_e^\mu \cdot \bar{x}_e = \sum_{e \in E} b_e \cdot \bar{x}_e$ (cf., e.g., (Ahuja et al., 1993)), we get that the flow $\bar{x}$ fulfills the following property:

$$\sum_{e \in E} d_e^{\pi,\mu} \cdot \bar{x}_e = \sum_{e \in E} \left(c_e^\pi - c_{\bar{e}}^\pi \cdot \frac{b_e^\mu}{b_{\bar{e}}^\mu} \right) \cdot \bar{x}_e = \sum_{e \in E} c_e^\pi \cdot \bar{x}_e - \frac{c_{\bar{e}}^\pi}{b_{\bar{e}}^\mu} \cdot \sum_{e \in E} b_e^\mu \cdot \bar{x}_e$$

$$= \sum_{e \in E} c_e \cdot \bar{x}_e - \frac{c_{\bar{e}}^\pi}{b_{\bar{e}}^\mu} \cdot \sum_{e \in E} b_e \cdot \bar{x}_e = \sum_{e \in E} c_e \cdot \bar{x}_e - \frac{c_{\bar{e}}^\pi}{b_{\bar{e}}^\mu} \cdot B. \tag{4.5}$$

Note that the value $-\frac{c_{\bar{e}}^\pi}{b_{\bar{e}}^\mu} \cdot B$ only depends on the basis structure and is independent from the flow $\bar{x}$. Hence, for any feasible flow $\bar{x}$ with $b(\bar{x}) = B$, the objective function values $d^{\pi,\mu}(\bar{x})$ and $c(\bar{x})$ only differ by a constant additive value, which shows the claim. □

Theorem 4.10:
For a feasible basis structure $(L, T, U, \bar{e})$ and the corresponding node potentials π and μ, assume that the reduced costs $d^{\pi,\mu}$ fulfill the following conditions:

$$d_e^{\pi,\mu} \geqslant 0 \text{ for all } e \in L, \tag{4.6a}$$
$$d_e^{\pi,\mu} = 0 \text{ for all } e \in T \cup \{\bar{e}\}, \tag{4.6b}$$
$$d_e^{\pi,\mu} \leqslant 0 \text{ for all } e \in U. \tag{4.6c}$$

Then the corresponding basic feasible flow x^* is optimal.

Proof: Let $d^{\pi,\mu}$ and x^* be defined as above and let $\bar{x}$ denote some arbitrary feasible flow. As shown in Lemma 4.9, minimizing $c(\bar{x}) = \sum_{e \in E} c_e \cdot \bar{x}_e$ is equivalent to minimizing $d^{\pi,\mu}(\bar{x}) = \sum_{e \in E} d_e^{\pi,\mu} \cdot \bar{x}_e$. Since we have

$$d^{\pi,\mu}(\bar{x}) = \sum_{e \in E} d_e^{\pi,\mu} \cdot \bar{x}_e = \sum_{e \in L} d_e^{\pi,\mu} \cdot \bar{x}_e + \sum_{e \in T \cup \{\bar{e}\}} d_e^{\pi,\mu} \cdot \bar{x}_e + \sum_{e \in U} d_e^{\pi,\mu} \cdot \bar{x}_e$$

$$= \sum_{e \in L} d_e^{\pi,\mu} \cdot \bar{x}_e + \sum_{e \in U} d_e^{\pi,\mu} \cdot \bar{x}_e \geqslant \sum_{e \in U} d_e^{\pi,\mu} \cdot u_e = d^{\pi,\mu}(x^*),$$

we get that x^* is optimal. □

4.3.4 Termination and Running Time

As shown above, we only make progress with respect to the objective function value if the corresponding simplex pivot is non-degenerate. However, like in the case of the traditional simplex method and the traditional network simplex algorithm (cf., e.g., (Ahuja et al., 1993; Dantzig, 1965)), it may be possible to end in an infinite loop of degenerate pivots if no further steps are undertaken. In the case of the traditional network simplex algorithm, there are two common methods to prevent cycling of the procedure: One can either use a perturbed problem, in which the right-hand side vector of the LP formulation is suitably transformed, or use the concept of *strongly feasible basis structures* in combination with a special leaving edge rule (cf. (Ahuja et al., 1988; Cunningham, 1976)). In this subsection, we show that a combination of both approaches leads to a finite network simplex algorithm with pseudo-polynomial running time for $BCMCFP_\mathbb{R}$.

In the following, we consider what we call the *transformed problem* of the given instance of $BCMCFP_\mathbb{R}$, in which we replace the (previously integral) budget B by $B' := B + \frac{1}{2}$. In doing so, we maintain feasibility of the problem: According to Assumption 4.5, the minimum cost flow x obtained by some minimum cost flow algorithm fulfills $b(x) \geqslant B + 1$. However, this implies that we can scale down x to a feasible flow x' with $b(x') = B'$, so we can restrict our considerations to the transformed problem. As it turns out, each basic feasible flow of the transformed problem fulfills a useful property that will be essential throughout this section:

Lemma 4.11:
For each basis structure $(L, T, U, \bar{e})$ of the transformed problem and its corresponding basic feasible flow x, it holds that $x_e \notin \mathbb{N}_{\geqslant 0}$ for all $e \in C(\bar{e})$.

Proof: According to Corollary 4.6, the flow x can be decomposed into an integral flow x^I and a flow x^C that is positive only on $C(\bar{e})$. Since $b_e \in \mathbb{N}_{\geqslant 0}$ for each $e \in E$, it holds that $\sum_{e \in E} b_e \cdot x_e^I \in \mathbb{N}_{\geqslant 0}$, so $b(x) = B' = B + \frac{1}{2}$ implies that $\sum_{e \in C(\bar{e})} b_e \cdot x_e^C = k + \frac{1}{2}$ for some integer k. Since x^C is positive only on $C(\bar{e})$, it holds that $x_e^C = \lambda$ for each $e \in C^+(\bar{e})$ and $x_e^C = -\lambda$ for each $e \in C^-(\bar{e})$ with $\lambda = \frac{k+\frac{1}{2}}{b(C(\bar{e}))} \notin \mathbb{N}_{\geqslant 0}$, which shows the claim. $\square$

We now show that we can restrict our considerations solely to the transformed problem since an optimal basic solution that is obtained by an application of the network simplex algorithm to the transformed problem also yields an optimal basic solution of the original problem:

Lemma 4.12:
An optimal basic solution of the transformed problem can be turned into an optimal basic solution of the original problem in $\mathcal{O}(n)$ time.

Proof: Let $(L, T, U, \bar{e})$ denote a basis structure that implies an optimal solution x^* of the transformed problem. According to Lemma 4.11, the flow x_e^* on each edge $e \in C(\bar{e})$ fulfills $x_e^* \notin \mathbb{N}_{\geqslant 0}$. In particular, this implies that $x_e^* \in (0, u_e)$ for each $e \in C(\bar{e})$, so we can increase or reduce the flow on the cycle by a small amount without violating any flow bounds. Since $x_e^* \in \mathbb{N}_{\geqslant 0}$ for each $e \notin C(\bar{e})$ according to Corollary 4.6 and since $b_e \in \mathbb{N}_{\geqslant 0}$ for each $e \in E$, it must hold that we can increase or reduce the flow on $C(\bar{e})$ by δ units such that $\delta \cdot b(C(\bar{e})) = -\frac{1}{2}$, i.e., such that we obtain a feasible flow for the original problem. Moreover, note that $d_e^{\pi,\mu} = 0$ for each $e \in C(\bar{e})$, i.e., the flow still fulfills the optimality conditions from Lemma 4.9. Since the flow values on the edges in $L \cup U$ remain unchanged, the resulting flow is the basic solution corresponding to the basis structure $(L, T, U, \bar{e})$ for the original problem, which shows the claim. $\square$

As noted above, one method to prevent cycling of the traditional network simplex algorithm is to use the concept of *strongly feasible basis structures*, which are feasible basis structures in which every empty tree edge is directed towards the root node and every full tree edge heads away from the root node (cf. (Ahuja et al., 1993)). As shown by Ahuja et al. (1988), an equivalent definition is that, in the corresponding basic feasible flow, it is possible to send a positive amount of additional flow from every node $v \in V$ to the root node via tree edges. For BCMCFP$_\mathbb{R}$, it turns out that a strongly feasible basis structure remains strongly feasible after a simplex pivot if the leaving edge is chosen appropriately, just as in the case of the traditional network simplex algorithm:

Lemma 4.13:
Let $(L, T, U, \bar{e})$ denote a strongly feasible basis structure of the transformed problem. The leaving edge e' can be chosen such that the basis structure $(L', T', U', \bar{e}')$ that results from a simplex pivot is again strongly feasible.

Proof: Let $e = (v, w)$ denote the entering edge (we assume that $e \in L$; the case that $e \in U$ works analogously) and let $E' \subseteq T \cup \{\bar{e}\}$ denote the set of blocking edges that determine the value of δ in the simplex pivot. Note that the graph that is induced by $T \cup \{\bar{e}, e\}$ contains up to three simple cycles, one of which must carry a fractional amount of flow after the simplex pivot according to Lemma 4.11. Hence, since the cycle that carries a fractional amount of flow cannot contain a blocking edge, there are three cases to distinguish: It either holds that $E' \subseteq C(e) \setminus C(\bar{e})$ or that $E' \subseteq C(\bar{e}) \setminus C(e)$ or that $E' \subseteq C(e) \cap C(\bar{e})$ (cf. Figure 4.2). We distinguish these three cases in the following. Note that, as in the proof of Lemma 4.8, we get that $C(e) \cap C(\bar{e})$ corresponds to a single simple path P_0 consisting of edges in T.

First assume that $E' \subseteq C(e) \setminus C(\bar{e})$. In this case, it holds that $\bar{e}' = \bar{e}$ and the cycle $C(\bar{e})$ continues to exist in $T' \cup \{\bar{e}'\}$. Hence, the flow on all edges in $C(\bar{e})$ remains fractional and we are still able to send a positive amount of flow from any node in $C(\bar{e})$ to

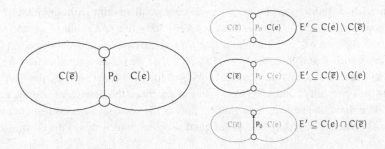

Figure 4.2: The graph induced by $T \cup \{e, \bar{e}\}$ (left) and the three possible cases for the set E' of blocking edges (right). In each of these cases, the set E' is contained in the solid black segment.

the apex[3] of $C(\bar{e})$. The rest of the proof of this case is analogous to the one for the traditional network simplex algorithm (cf., e.g., (Ahuja et al., 1993) and Figure 4.3): We choose the leaving edge $e' = (v', w')$ to be the last edge in E' that occurs when traversing the cycle $C(e)$ in the direction of e, starting at the apex z of $C(e)$. For the sake of notational simplicity, assume that $e' \in C^+(e)$ (the case that $e' \in C^-(e)$ works analogously). Clearly, if $\bar{v}$ is any node on the path from w' to z (in the direction of the cycle), we can still send a positive amount of flow from $\bar{v}$ to z since there are no further blocking edges on this path according to the choice of e'. On the other hand, for the case that the simplex pivot is non-degenerate, we send a positive amount of flow on the path from z to v' (which may be reduced by the flow on the cycle $C(\bar{e})$ on the edges in $C(e) \cap C(\bar{e})$, but which will not be reduced to zero since $C(\bar{e})$ contains no blocking edges). Hence, we can send a positive amount of flow back from every node $\bar{v}$ on the path from v' to z in $T' \cup \{\bar{e}'\} = T \cup \{e, \bar{e}\} \setminus \{e'\}$. For the case that the simplex pivot is degenerate, it must hold that all blocking edges E' lie on the path from z to v since $(L, T, U, \bar{e})$ is strongly feasible and we can, thus, send a positive amount of flow to z on every edge on the path from w to z. However, in the degenerate case, we do not change the flow on the edges on the path from z to v' and can, thus, still send a positive amount of flow from v' to z. So, in any case, we can still send flow from any node on $C(e)$ after removing edge e'. Note that the flow does not change on any edge in $E \setminus (C(e) \cup C(\bar{e}))$, so we can send a positive amount of flow from every node in V to the root node r after the simplex pivot (possibly via the edges in $C(e) \setminus \{e'\}$). Hence, we maintain a strongly feasible basis structure in this case.

---

3   The *apex* of a cycle $C(e)$ with $e = (v, w)$ is the first common node of the two unique paths in T from v to the root and from w to the root.

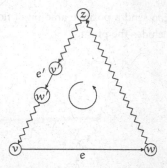

Figure 4.3: A cycle $C(e)$ that is induced by the entering edge $e = (v, w) \in L$ with $d_e^{\pi,\mu} < 0$. After sending flow on $C(e)$, the leaving edge e' is the last blocking edge when traversing $C(e)$ in the direction of e starting from the apex z.

The second case, in which $E' \subseteq C(\overline{e}) \setminus C(e)$, works similar to the previous case. Note that we now get that $\overline{e}' = e$ since the cycle $C(\overline{e})$ vanishes. We choose the leaving edge to be the last blocking edge that occurs when traversing the cycle $C(\overline{e})$ in the direction in that we send the flow in the simplex pivot, starting at the apex of $C(\overline{e})$. Note that the simplex pivot must be non-degenerate in this case since the blocking edges lie on $C(\overline{e})$ and every edge on $C(\overline{e})$ carries a fractional amount of flow before the simplex pivot.

It remains to show that we maintain a strongly feasible basis structure in the case that $E' \subseteq P_0 = C(e) \cap C(\overline{e})$ (cf. Figure 4.4). As in the previous case, the simplex pivot is non-degenerate since all edges in $C(\overline{e})$ carry a fractional amount of flow before the simplex pivot. Thus, the algorithm sends a positive amount of flow δ along $C(e)$ and $\overline{\delta} := -\frac{b(C(e))}{b(C(\overline{e}))} = -\frac{b_e^\mu}{b_{\overline{e}}}$ units of flow along $C(\overline{e})$. Since no edge in $C(e) \setminus C(\overline{e})$ and $C(\overline{e}) \setminus C(e)$ is a blocking edge and we send flow on both cycles, neither of these edges is empty or full after the simplex pivot (this also follows from the fact that the new cycle in $T' \cup \{\overline{e}'\} = T \cup \{\overline{e}, e\} \setminus \{e\}$ consists of the edges in $(C(e) \setminus C(\overline{e})) \cup (C(\overline{e}) \setminus C(e))$ and, thus, carries a fractional amount of flow, cf. the gray paths in Figure 4.4). Moreover, since $E' \subseteq P_0$ (and $E' \neq \emptyset$), we must have $\delta \neq \overline{\delta}$. Thus, there is a unique direction in which the flow is sent on P_0 (from z to $\overline{w}$ in Figure 4.4). We choose the leaving edge $e' = (v', w')$ to be the last blocking edge that occurs on any of the two cycles when traversing the corresponding cycle in the direction of this flow, starting from the apex of the cycle. We can then send flow from w' to the apexes of both cycles (since there are no further blocking edges on the corresponding subpath of P_0 and since the flow on the new cycle is fractional) and from v' to the apexes (since we have sent a positive amount of flow to v' on the corresponding subpath of P_0 and since the flow on the new cycle is fractional). Hence, using the same arguments as in the previous two cases, we get that we can send a positive amount of flow from each node $v \in V$ to the root node, which concludes the proof. □

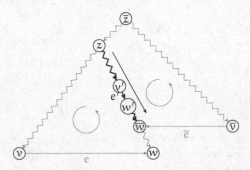

Figure 4.4: The case that $E' \subseteq P_0 = C(e) \cap C(\bar{e})$. The leaving edge e' is chosen to be the last edge on P_0 when traversing any of the two cycles $C(e)$ and $C(\bar{e})$ in the direction of the flow change on P_0.

Lemma 4.13 builds the foundation for the following theorem, which shows that the network simplex algorithm for BCMCFP$_\mathbb{R}$ does not cycle when using strongly feasible basis structures:

Theorem 4.14:
The network simplex algorithm applied to the transformed problem terminates within a finite number of simplex pivots when using strongly feasible basis structures. Moreover, the number of consecutive degenerate simplex pivots is bounded by $\mathcal{O}(n^3 CB)$.

Proof: Let $(L, T, U, \bar{e})$ denote a basis structure and let x denote its corresponding basic feasible flow. Consider a degenerate simplex pivot that occurs when adding the entering edge $e = (v, w)$ to $T \cup \{\bar{e}\}$ and choosing the leaving edge $e' = (v', w')$ according to the leaving edge rules given in the proof of Lemma 4.13, which leads to a new basis structure $(L', T', U', \bar{e}')$. Since the flow on $C(\bar{e})$ is fractional according to Lemma 4.11, none of the edges on the cycle $C(\bar{e})$ can be degenerate. Thus, it holds that $\bar{e}' = \bar{e}$ and that the cycle $C(\bar{e})$ continues to exist in $T' \cup \{\bar{e}'\}$. Moreover, as in the proof of Lemma 4.13, the leaving edge e' must lie on the path in T from the apex z of $C(e)$ to v since the basis structure is strongly feasible (cf. Figure 4.3). As in the traditional network simplex algorithm, after the simplex pivot, the node potentials π_v and μ_v are increased (decreased) by c_e^π and b_e^μ, respectively, for all nodes v on the path from w' to v in T if $e \in L$ ($e \in U$), while the remaining node potentials remain unchanged (cf., e.g., (Ahuja et al., 1993)). Thus, for each v on this path, the new node potentials π'_v and μ'_v fulfill

$$\pi'_v + \frac{c_{\bar{e}}^\pi}{b_{\bar{e}}^\mu} \cdot \mu'_v = (\pi_v + c_e^\pi) + \frac{c_{\bar{e}}^\pi}{b_{\bar{e}}^\mu} \cdot (\mu_v + b_e^\mu)$$

$$= \left(\pi_v + \frac{c_{\bar{e}}^\pi}{b_{\bar{e}}^\mu} \cdot \mu_v\right) + \left(c_e^\pi + \frac{c_{\bar{e}}^\pi}{b_{\bar{e}}^\mu} \cdot b_e^\mu\right)$$

$$= \left(\pi_v + \frac{c_{\bar{e}}^\pi}{b_{\bar{e}}^\mu} \cdot \mu_v\right) + d_e^{\pi,\mu} < \pi_v + \frac{c_{\bar{e}}^\pi}{b_{\bar{e}}^\mu} \cdot \mu_v$$

for the case that $e \in L$. Otherwise, if $e \in U$, we get that

$$\pi'_v + \frac{c_{\bar{e}}^\pi}{b_{\bar{e}}^\mu} \cdot \mu'_v = (\pi_v - c_e^\pi) + \frac{c_{\bar{e}}^\pi}{b_{\bar{e}}^\mu} \cdot (\mu_v - b_e^\mu)$$

$$= \left(\pi_v + \frac{c_{\bar{e}}^\pi}{b_{\bar{e}}^\mu} \cdot \mu_v\right) - \left(c_e^\pi + \frac{c_{\bar{e}}^\pi}{b_{\bar{e}}^\mu} \cdot b_e^\mu\right)$$

$$= \left(\pi_v + \frac{c_{\bar{e}}^\pi}{b_{\bar{e}}^\mu} \cdot \mu_v\right) - d_e^{\pi,\mu} < \pi_v + \frac{c_{\bar{e}}^\pi}{b_{\bar{e}}^\mu} \cdot \mu_v.$$

Hence, the value $\sum_{v \in V} \pi_v + \frac{c_{\bar{e}}^\pi}{b_{\bar{e}}^\mu} \cdot \mu_v$ decreases strictly after each degenerate simplex pivot. However, the values π_v are integers in $\{-nC, \ldots, nC\}$, while the values μ_v lie in $\{-nB, \ldots, nB\}$ for each $v \in V$ (cf. (Ahuja et al., 1993)). Thus, since there are only $\mathcal{O}(n^2 CB)$ combinations of integral values for π_v and μ_v for each node $v \in V$ and since the fraction $\frac{c_{\bar{e}}^\pi}{b_{\bar{e}}^\mu}$ remains unchanged during degenerate pivots, the algorithm performs a non-degenerate pivot after at most $\mathcal{O}(n^3 CB)$ degenerate simplex pivots and the claim follows. □

While the leaving edge rules as described above guarantee finiteness of the procedure, we can reduce the number of non-degenerate simplex pivots by applying Dantzig's pivoting rule (cf., e.g., (Ahuja et al., 1988)), i.e., by choosing the entering edge to be the one with the largest violation of its optimality condition:

Lemma 4.15:
The network simplex algorithm applied to the transformed problem performs a total number of $\mathcal{O}(nmUB \log(mCUB))$ non-degenerate simplex pivots when using Dantzig's pivoting rule.

Proof: The proof of the lemma is similar to the one for the traditional network simplex algorithm given in (Ahuja et al., 1988). Let x^k denote the basic feasible flow that is obtained after the k-th non-degenerate step of the algorithm and let $c(x^k)$ denote its objective function value. Moreover, let $(L, T, U, \bar{e})$ denote the corresponding basis structure. According to Corollary 4.6, each flow x^k can be decomposed into an integral flow x^I and a fractional flow x^C on the cycle $C(\bar{e})$. In particular, since x^k satisfies $b(x^k) = B' = B + \frac{1}{2}$ and x^C is a flow on the cycle $C(\bar{e})$, it holds that $x_e^C = \frac{p}{b(C(\bar{e}))}$ for $p := (B + \frac{1}{2}) - \sum_{e \in E} b_e \cdot x_e^I$, so the flow on every edge in x^k is an integral multiple of $\frac{1}{2 \cdot b(C(\bar{e}))}$. Thus, it holds that $|x_e^k - x_e^{k+1}|$ is either zero or at least $\frac{1}{2 \cdot b(C(\bar{e}))} \geq \frac{1}{2nB}$ for each edge $e \in E$. Moreover, note that the minimum absolute value of the reduced costs of any edge e that violates its optimality condition can be bounded as follows:

$$|d_e^{\pi,\mu}| = \left| c_e^\pi - \frac{b_e^\mu}{b_{\bar{e}}^\mu} \cdot c_{\bar{e}}^\pi \right| = \left| \frac{c_e^\pi \cdot b_{\bar{e}}^\mu - b_e^\mu \cdot c_{\bar{e}}^\pi}{b_{\bar{e}}^\mu} \right| \geq \frac{1}{b(C(\bar{e}))} \geq \frac{1}{nB}.$$

Since the objective function value of any flow is bounded from below by $-mCU$, we, thus, get that the maximum number of non-degenerate simplex pivots *without* using Dantzig's pivoting rule is bounded by $\mathcal{O}(mCU \cdot 2nB \cdot nB) = \mathcal{O}(n^2 mCUB^2)$.

Let $\Delta := \max\{-\min_{e \in L} d_e^{\pi,\mu}, \max_{e \in U} d_e^{\pi,\mu}\}$ denote the maximum violation of the optimality conditions of any edge in $L \cup U$ and let e denote the corresponding edge that is chosen based on Dantzig's pivoting rule. Since sending one unit of flow over $C(e)$ reduces the objective function value by Δ, we get that

$$c(x^k) - c(x^{k+1}) \geq \frac{\Delta}{2nB} \tag{4.7}$$

Moreover, if x^* denotes an optimal solution to the problem, we get according to equation (4.5) in the proof of Lemma 4.9 that

$$c(x^k) - c(x^*) = d^{\pi,\mu}(x^k) - d^{\pi,\mu}(x^*) = \sum_{e \in E} d_e^{\pi,\mu} \cdot (x_e^k - x_e^*)$$

$$= \sum_{e \in L} d_e^{\pi,\mu} \cdot (-x_e^*) + \sum_{e \in U} d_e^{\pi,\mu} \cdot (u_e - x_e^*) \leq m\Delta U. \tag{4.8}$$

Combining equations (4.7) and (4.8), we, thus, get that

$$c(x^k) - c(x^{k+1}) \geq \frac{c(x^k) - c(x^*)}{2nmUB},$$

i.e., after each non-degenerate simplex pivot, the gap to the optimal solution with respect to the objective function value is reduced by a factor of at least $\frac{1}{2nmUB}$. Ahuja et al. (1993) show that, if H is the maximum number of improving steps of any algorithm and if this algorithm reduces the gap to the optimal solution by a fraction of at least α in each step, then the maximum number of steps is bounded by $\mathcal{O}(\frac{1}{\alpha} \log H)$. Thus, since $H \in \mathcal{O}(n^2 mCUB^2)$ in our case as shown above, we get that the maximum number of non-degenerate simplex pivots using Dantzig's pivoting rule is in $\mathcal{O}(2nmUB \log(n^2 mCUB^2)) = \mathcal{O}(nmUB \log(mCUB))$. $\square$

In Lemma 4.13, it was shown that we can obtain a strongly feasible basis structure again when performing a simplex pivot on a strongly feasible basis structure. However, as a last ingredient, it remains open how to determine an initial strongly feasible basis structure. This will be shown in the following lemma:

Lemma 4.16:
An initial strongly feasible basis structure $(L, T, U, \bar{e})$ for BCMCFP$_\mathbb{R}$ and the corresponding basic feasible solution x can be determined in $\mathcal{O}(m)$ time.

Proof: Similar to (Çalışkan, 2011), we insert an artificial edge $e_0 = (s, t)$ with costs $c_{e_0} := 1$, capacity $u_{e_0} := 1$, and usage fees $b_{e_0} := 2B + 1 = 2B'$. The initial basic feasible solution x is defined by $x_{e_0} := 0.5$ and $x_e := 0$ for each $e \in E$. The spanning tree T consists of e_0 as well as a spanning tree of the nodes in $V \setminus \{s\}$ that is a directed in-tree with root t. Note that such an in-tree exists according to Assumption 4.3 and can be found, e.g., by a depth-first search in $\mathcal{O}(m)$ time. Moreover, note that we can send a positive amount of flow from every node in V to s by using the unique path in the in-tree in combination with e_0. Hence, we obtain a strongly feasible basis structure by setting $\bar{e} := e$ for some $e \in \delta^+(s) \setminus \{e_0\}$, choosing T as defined above, and setting $U := \emptyset$ and $L := E \setminus (T \cup \{\bar{e}\})$. Note that the new edge neither influences the bounds C, U, and B (since $c_{e_0} \in \mathcal{O}(C)$, $u_{e_0} \in \mathcal{O}(U)$, and $b_{e_0} \in \mathcal{O}(B)$ in any instance) nor the optimal solution (since e_0 will be empty in any optimal solution). $\square$

We are now ready to prove the main result of the section:

Theorem 4.17:
The network simplex algorithm for BCMCFP$_\mathbb{R}$ can be implemented to run in $\mathcal{O}(n^4 m^2 \cdot CUB^2 \cdot \log(mCUB))$ time.

Proof: According to Lemma 4.16, we can determine an initial basis structure and the corresponding basic feasible solution in $\mathcal{O}(m)$ time. It is easy to see that a single simplex pivot as described above can be implemented to run in $\mathcal{O}(m)$ time, including the overhead to determine the entering edge and the leaving edge according to Dantzig's pivoting rule and the above leaving edge rules. Moreover, the maximum number of non-degenerate simplex pivots is given by $\mathcal{O}(nmUB \log(mCUB))$ as shown in Lemma 4.15. In the worst case, each of these non-degenerate simplex pivots is followed by a sequence of $\mathcal{O}(n^3 BC)$ degenerate pivots according to Theorem 4.14, which leads to an overall running time of

$$\mathcal{O}(m \cdot n^3 BC \cdot nmUB \log(mCUB)) = \mathcal{O}(n^4 m^2 CUB^2 \log(mCUB)),$$

which shows the claim. $\square$

As a summary, we were able to adapt the basic underlying ideas used in the traditional network simplex algorithm to the case of BCMCFP$_\mathbb{R}$. In particular, we generalized the notions of basis structures, reduced costs, and simplex pivots and managed to apply to common techniques to avoid cycling. As a result, we obtained a fully combinatorial network simplex algorithm for the budget-constrained minimum cost flow problem. The theoretical time-bound of this algorithm is worse than the one of the network simplex algorithm for the traditional minimum cost flow problem and worse than the one of the algorithms presented in the upcoming section. Nevertheless, as it is usual for network simplex variants, empirical results suggest that the underlying structure leads to a superior empirical performance in comparison to other solution methods (cf. (Çalışkan, 2011)).

4.4 Bicriteria Interpretation

In this section, we present a very different approach to solve $\text{BCMCFP}_\mathbb{R}$ that relies on an interpretation of the problem as a bicriteria minimum cost flow problem. In doing so, we obtain a weakly polynomial-time algorithm based on a binary search. Moreover, by incorporating Megiddo's parametric search technique as introduced in Section 3.1, we show that the problem becomes solvable in strongly polynomial time.

4.4.1 A Weakly Polynomial-Time Algorithm

The fact that we are dealing with two kinds of costs in the case of $\text{BCMCFP}_\mathbb{R}$ leads to the impression that the problem may be somehow related to the bicriteria minimum cost flow problem. As noted above, the problem $\text{BCMCFP}_\mathbb{R}$ can be interpreted as the ε-constraint method applied to the bicriteria minimum cost flow problem (cf. (Chankong and Haimes, 2008)). This fact can be seen as follows: Since we seek to minimize the cost function c while maintaining $b(x) \leqslant B$, each optimal solution x^* of $\text{BCMCFP}_\mathbb{R}$ corresponds to a point $(c(x^*), b(x^*))^\top$ in the objective space that lies on the pareto frontier and not above the line $b = B$. In fact, according to Assumption 4.5 and the results of the previous section, we can assume the optimal solution to fulfill $b(x^*) = B$, i.e., the desired solution lies on the line $b = B$ in the objective space. The situation is shown in Figure 4.5.

It is well-known that, for each point $(c, b)^\top$ on the pareto frontier, there is some value $\lambda \in [0, \infty)$ and a feasible flow x with $(c(x), b(x))^\top = (c, b)^\top$ such that x is a minimum cost flow with respect to the costs $b_e + \lambda \cdot c_e$ for each edge $e \in E$ (cf. Geoffrion (1967); Ehrgott (2005)). Assume that there are two flows $x^{(1)}$ and $x^{(2)}$ which are both optimal for some specific value of λ, i.e., $b(x^{(1)}) + \lambda \cdot c(x^{(1)}) = b(x^{(2)}) + \lambda \cdot c(x^{(2)}) = \alpha$ for some value α. Then, for both of the flows $x^{(i)}$ with $i \in \{1, 2\}$, it holds that $b(x^{(i)}) = \alpha - \lambda \cdot c(x^{(i)})$, i.e., they lie on the same *efficient edge*, which is a straight

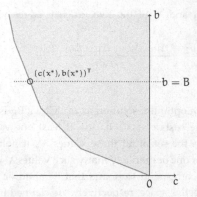

Figure 4.5: The objective space of the interpretation of BCMCFP$_{\mathbb{R}}$ as a bicriteria minimum cost flow problem. The orange area corresponds to the set of the objective values of feasible flows. The thick orange lines correspond to efficient edges, which form the pareto frontier.

line with slope $-\lambda$ in the objective space. In other words, computing a minimum cost flow with edge costs $b_e + \lambda \cdot c_e$ will either provide a solution that lies on an extreme point of the pareto frontier or some point that lies on the efficient edge with slope $-\lambda$. Moreover, the slopes of these efficient edges differ by a minimum absolute amount, which is shown in the following lemma:

Lemma 4.18:
The slopes of two efficient edges on the pareto frontier of any instance of BCMCFP$_{\mathbb{R}}$ differ by an absolute value of at least $\frac{1}{\bar{c}^2}$ for $\bar{c} := \sum_{e \in E} |u_e \cdot c_e|$.

Proof: Consider some extreme point (c, b) of the pareto frontier. The pair $(c, b)^\mathsf{T} = (c(x), b(x))^\mathsf{T}$ is the (bicriteria) objective value of a flow x that is a (single objective) minimum cost flow with respect to costs $b_e + \lambda \cdot c_e$ for some specific value of λ. Since there is always an integral minimum cost flow if the capacities are integral (cf. Section 2.4), we may assume that both c and b are integral as well. Therefore, the slope of each efficient edge can be described by the fraction $\frac{\Delta b}{\Delta c}$ of two integers Δb and Δc. Let $\lambda_1 := \frac{\Delta b_1}{\Delta c_1}$ and $\lambda_2 := \frac{\Delta b_2}{\Delta c_2}$ be the slopes of two efficient edges γ_1 and γ_2, respectively. Without loss of generality, we assume that $\lambda_1 > \lambda_2$, so we obtain

$$\lambda_1 > \lambda_2 \Longleftrightarrow \frac{\Delta b_1}{\Delta c_1} > \frac{\Delta b_2}{\Delta c_2} \Longleftrightarrow \frac{\Delta b_1}{\Delta c_1} - \frac{\Delta b_2}{\Delta c_2} > 0$$
$$\Longleftrightarrow \Delta b_1 \cdot \Delta c_2 - \Delta b_2 \cdot \Delta c_1 > 0 \Longleftrightarrow \Delta b_1 \cdot \Delta c_2 - \Delta b_2 \cdot \Delta c_1 \geqslant 1.$$

Therefore, the slopes λ_1 and λ_2 of the two efficient edges γ_1 and γ_2 differ by an absolute value of

$$\lambda_1 - \lambda_2 = \frac{\Delta b_1}{\Delta c_1} - \frac{\Delta b_2}{\Delta c_2} = \frac{\Delta b_1 \cdot \Delta c_2 - \Delta b_2 \cdot \Delta c_1}{\Delta c_1 \cdot \Delta c_2} \geqslant \frac{1}{\bar{c}^2}$$

as claimed. □

As explained above, each optimum solution x^* of $BCMCFP_\mathbb{R}$ is a minimum cost flow with respect to the edge costs $c_e + \lambda^* \cdot b_e$ for at least one value $\lambda^* \in [0, +\infty)$. In particular, if Λ^* denotes the set of all these values λ^*, it holds that Λ^* is a closed interval containing either one or infinitely many such values λ^* depending on whether the optimum solutions correspond to points that lie amid or at the corner of some efficient edge in the objective space, respectively. As claimed in the following lemma we are able to decide the membership in Λ^* efficiently:

Lemma 4.19:
Let $\Lambda^* \neq \emptyset$ denote the set of parameters λ^* for which an optimum solution x^* to $BCMCFP_\mathbb{R}$ is a minimum cost flow with respect to the edge costs $c_e + \lambda^* \cdot b_e$ for each $e \in E$. For some candidate value λ, it is possible to decide whether $\lambda < \min \Lambda^*$, $\lambda > \max \Lambda^*$, or $\lambda \in \Lambda^*$ in $\mathcal{O}(MCF(m, n, C, U))$ and $\mathcal{O}(MCF(m, n))$ time.

Proof: Let x be the solution of a minimum cost flow computation with edge costs $b_e + \lambda \cdot c_e$ for each $e \in E$. Clearly, if $b(x) = B$, we have found an optimal solution of $BCMCFP_\mathbb{R}$ and are done. Now assume that $b(x) < B$ (the case that $b(x) > B$ is symmetric). Obviously, there are two possible cases: Either $\lambda < \min \Lambda^*$ (cf. Figure 4.6a) or $\lambda \in \Lambda^*$ but there are several minimum cost flows x' on the same efficient edge some of which fulfill $b(x') < B$ and some of which fulfill $b(x') \geqslant B$ (cf. Figure 4.6b). We claim that we can distinguish these two cases by computing two additional minimum cost flows x^+ and x^- using the costs $b_e + (\lambda + \delta) \cdot c_e$ and $b_e + (\lambda - \delta) \cdot c_e$ for $\delta := \frac{1}{2\bar{c}^2}$, respectively: Assume that at least two of the points $(c(x^+), b(x^+))^\mathsf{T}$, $(c(x), b(x))^\mathsf{T}$, and $(c(x^-), b(x^-))^\mathsf{T}$ coincide. Then x must lie on an extreme point of the efficient frontier (cf. Figure 4.6a) and, since $b(x) < B$, we have that $\lambda < \min \Lambda^*$. On the other hand, if the three flows have distinct objective values (cf. Figure 4.6b), the point $(c(x), b(x))^\mathsf{T}$ must lie amid some efficient edge γ with the extreme points $(c(x^+), b(x^+))^\mathsf{T}$ and $(c(x^-), b(x^-))^\mathsf{T}$ since the slope of two adjacent efficient edges differs by an absolute value of at least 2δ as seen in Lemma 4.18. Hence, the points $(c(x^+), b(x^+))^\mathsf{T}$ and $(c(x^-), b(x^-))^\mathsf{T}$ lie on the same efficient edge and we have $\lambda < \min \Lambda^*$ as well if $b(x^+) < B$. Otherwise, $\lambda \in \Lambda^*$ and the minimum cost flow retrieves an optimum solution to $BCMCFP_\mathbb{R}$. Note that the running time of this procedure is dominated by the three minimum cost flow computations and, hence, lies both in $\mathcal{O}(MCF(m, n))$ and $\mathcal{O}(MCF(m, n, C, U))$. □

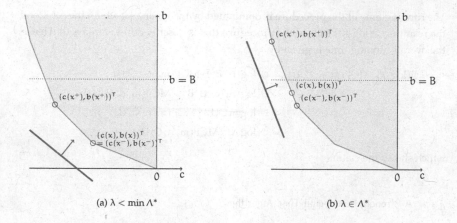

Figure 4.6: The procedure for comparing a candidate value for λ to the set of optimal values Λ^*. In the left (right) figure, the situation for a too small (correct) value for λ is shown.

We now show how we can use Lemma 4.19 within a binary search in order to obtain a weakly polynomial-time algorithm that performs within a factor $\mathcal{O}(\log M)$ of each algorithm for the traditional minimum cost flow problem:

Theorem 4.20:
$\text{BCMCFP}_\mathbb{R}$ is solvable in weakly polynomial time $\mathcal{O}(\log M \cdot \text{MCF}(m, n, C, U))$.

Proof: Consider the set $K := \left\{ k \cdot \frac{1}{2\overline{c}^2} : k \in \{0, \ldots, \overline{b} \cdot 2\overline{c}^2\} \right\}$, where $\overline{b} := \sum_{e \in E} u_e \cdot b_e \in \mathcal{O}(mUB)$ and $\overline{c} := \sum_{e \in E} |u_e \cdot c_e| \in \mathcal{O}(mCU)$ are upper bounds on the total usage fees and total absolute value of the costs of any feasible flow, respectively. Note that each extreme point of the pareto frontier can be obtained by a minimum cost flow computation with edge costs $c_e + \lambda \cdot b_e$ for some $\lambda \in K$ since the slopes of any two efficient edges differ by an absolute amount of at least $\frac{1}{\overline{c}^2}$ according to Lemma 4.18. Hence, by incorporating the procedure that is described in Lemma 4.19 into a binary search on the set K, we either find some value $\lambda \in \Lambda^*$ (in which case we have also found an optimum solution to $\text{BCMCFP}_\mathbb{R}$) or two "adjacent" values $\lambda^{(1)} := k \cdot \frac{1}{2\overline{c}^2}$ and $\lambda^{(2)} := (k+1) \cdot \frac{1}{2\overline{c}^2}$ for some $k \in \{0, \ldots, \overline{b} \cdot 2\overline{c}^2 - 1\}$ with $\lambda^{(1)} < \min \Lambda^*$ and $\lambda^{(2)} > \max \Lambda^*$. These values, however, yield solutions $x^{(1)}$ and $x^{(2)}$ that correspond to the corner points of the same efficient edge, which crosses the line $b = B$ in the objective space. Thus, by computing a suitable convex combination of the two solutions $x^{(1)}$ and $x^{(2)}$, we obtain an optimum solution to $\text{BCMCFP}_\mathbb{R}$ and are done.

The running time of the procedure is dominated by the binary search on the set K and the resulting $\mathcal{O}(\log|K|)$ calls to the procedure that is described in Lemma 4.19. Hence, the overall running time is given by

$$
\begin{aligned}
\mathcal{O}(\log|K| \cdot \text{MCF}(m, n, C, U)) &= \mathcal{O}(\log(\overline{b} \cdot \overline{c}^2) \cdot \text{MCF}(m, n, C, U)) \\
&= \mathcal{O}(\log(m^3 C^2 U^4 B^2) \cdot \text{MCF}(m, n, C, U)) \\
&= \mathcal{O}(\log mCUB \cdot \text{MCF}(m, n, C, U)) \\
&= \mathcal{O}(\log M \cdot \text{MCF}(m, n, C, U)),
\end{aligned}
$$

which shows the claim. □

4.4.2 A Strongly Polynomial-Time Algorithm

Besides an incorporation in a binary search, the result of Lemma 4.19 can also be used in combination with Megiddo's parametric search technique in order to determine a suitable value for the weighting parameter λ. In order to do so, we simulate Orlin's (1993) enhanced capacity scaling algorithm, which runs in $\mathcal{O}(m \log n \cdot (m + n \log n))$ time on *simple* graphs. As it will be shown in the following lemma, the running time of the algorithm worsens only slightly for the problem on multigraphs:

Lemma 4.21:
The enhanced capacity scaling algorithm can be implemented to run in $\mathcal{O}(m \log m \cdot (m + n \log n))$ time on multigraphs.

Proof: The enhanced capacity scaling algorithm as introduced in (Orlin, 1993) requires the edges to be uncapacitated. Therefore, in a preprocessing step, a graph with capacities on each edge is transformed into an equivalent graph as follows: For each edge $e \in E$ with finite capacity $u_e \in \mathbb{N}_{\geqslant 0}$ that heads from some node $u \in V$ to $v \in V$, we introduce an artificial node v' and replace e by two uncapacitated edges $e' = (u, v')$ with $c_{e'} = c_e$ and $e'' = (v, v')$ with $c_{e''} = 0$. Furthermore, we assign a demand of u_e to node v'. We refer to Ahuja et al. (1993) for further details on the transformation.

After the transformation, the graph does not contain parallel edges any more. However, the number of edges increases from m to $m' \leqslant 2m$ while the number of nodes increases from n to $n' \leqslant m + n$. By analyzing the complexity of the algorithm as done in Ahuja et al. (1993) *without* using that $m \in \mathcal{O}(n^2)$ (which is clearly no longer valid for multigraphs), we obtain a running time of $\mathcal{O}((m + n) \log(m + n) \cdot \text{SP}(2m, m + n)) = \mathcal{O}(m \log m \cdot \text{SP}(2m, m + n))$. Recall from Section 2.4 that $\text{SP}(m, n) = \mathcal{O}(m + n \log n)$ when using Dijkstra's algorithm in combination with Fibonaccci heaps. This leads to a running time of $\mathcal{O}(m \log m \cdot (2m + (m + n) \log(m + n))) = \mathcal{O}(m \log m \cdot (m + m \log m))$

for the enhanced capacity scaling algorithm. However, we can use the following observation to improve this running time: Whenever we update the distance label of an artificial node v' while exploring some adjacent node v during the execution of Dijkstra's algorithm, we can immediately update the label of the second node u that is adjacent to v' in constant time if there is an edge with positive residual capacity from v' to u. Thus, we do not need to handle node v' explicitly and the number of nodes that need to be inserted and extracted from the Fibonacci heap reduces to n. Hence, we get the claimed running time of $\mathcal{O}(m \log m \cdot (m + n \log n))$ for the enhanced capacity scaling algorithm on multigraphs. $\qquad\square$

Lemma 4.21 allows us to obtain a strongly polynomial-time algorithm for the budget-constrained minimum cost flow problem:

Theorem 4.22:
For $f(n) \in o(n^3)$, BCMCFP$_\mathbb{R}$ is solvable in strongly polynomial time $\tilde{\mathcal{O}}(\min\{nm \cdot \text{MCF}(m,n), m \cdot \text{MCF}(m,n) + m \cdot f(n)\})$.

Proof: The idea of the algorithm is to handle BCMCFP$_\mathbb{R}$ as a bicriteria minimum cost flow problem and to find a flow that satisfies the budget constraint (4.1c) using Megiddo's parametric search technique as introduced in Section 3.1. By directing the execution of the enhanced capacity scaling algorithm using Lemma 4.19, we finally come up with a solution that lies on the efficient edge crossing the line $b = B$ in the objective space or an extreme point that lies on this line. This will be shown in the following.

We simulate the enhanced capacity scaling algorithm with edge costs $a(\lambda) := b_e + \lambda \cdot c_e$ for each edge $e \in E$, where λ is kept as a symbolic variable. Moreover, we maintain an interval I, initialized by $I = [0, \infty)$, for which we know that $\Lambda^* \subseteq I$. As it can be easily seen, the enhanced capacity scaling algorithm is a strongly combinatorial algorithm according to the definition given in Section 3.1, so Megiddo's parametric search technique can be applied. By incorporating Lemma 4.19, we can determine if a candidate value λ' is smaller than $\min \Lambda^*$, larger than $\max \Lambda^*$, or contained in Λ^* in $\mathcal{O}(\text{MCF}(m,n))$ time. This fact allows us to resolve each comparison that may occur during the simulation of the algorithm and to update the interval I appropriately. At the end of the simulation, the resulting flow $x(\lambda)$ (which may still depend on λ) is optimal for every choice of $\lambda \in I$, but may only fulfill $b(x) = B$ for some specific value in I. Nevertheless, by solving the equation $b(x(\lambda)) = B$, we finally obtain an optimal value $\lambda^* \in \Lambda^*$ and a solution for BCMCFP$_\mathbb{R}$.

Without any further parallelization techniques, the above procedure would result in a time bound of $\mathcal{O}(m \log m \cdot (m + n \log n) \cdot \text{MCF}(m,n))$ according to Lemma 4.21 since we need to evaluate the callback for each of the $\mathcal{O}(m \log m \cdot (m + n \log n))$ steps of the algorithm in the worst case. We now show that the problem can be solved in $\mathcal{O}(m \log m \cdot \min\{T_1(m, n), T_2(m, n)\})$ time, where

$$T_1(m, n) \in \mathcal{O}\left((n + \log \log m) \log m \cdot \mathrm{MCF}(m, n) + m\right)$$

and, for $f(n) \in o(n^3)$,

$$T_2(m, n) \in \mathcal{O}\left(\left(\log^2 n \log \log n + \log m \log \log m\right) \cdot \mathrm{MCF}(m, n) + f(n)\right).$$

By ignoring poly-logarithmic factors, this leads to the claimed strongly polynomial running time of $\widetilde{\mathcal{O}}(\min\{nm \cdot \mathrm{MCF}(m, n), m \cdot \mathrm{MCF}(m, n) + m \cdot f(n)\})$.

In order to do so, we first need to briefly investigate the mechanics of the enhanced capacity scaling algorithm. The algorithm works in $\mathcal{O}(m \log m)$ phases and performs $\mathcal{O}(m \log m)$ shortest path augmentations. In every phase of the algorithm, each edge of the graph is either labeled as *abundant* or *non-abundant*. The subgraph that is induced by the abundant edges is called the *abundant subgraph* and contains several connected components, called *abundant components*. At the beginning of the algorithm, each node of the graph is considered as one distinct abundant component. Moreover, each abundant component contains one distinguished node, called *root node*. The imbalance must be zero for each node except for the roots of the abundant components.

In each phase, the algorithm performs the following operations on the residual network based on the current flow:

1. It checks whether there are nodes with positive imbalance.

2. If so, it checks whether the imbalance of each node is less or equal to some constant value.

3. For each edge e, it checks if the flow value x_e on e is greater or equal to some constant value.

4. If the above is true for some edge e, the edge is marked as abundant and the two components of the abundant subgraph that are now connected by e, say C_1 with root node v_1 and C_2 with root node v_2, become merged into one component. Moreover, if q is the imbalance at v_2, we send q units of flow from v_2 to v_1 using abundant edges only.

During the simulation of the enhanced capacity scaling algorithm, we are able to perform each of the above steps in $\mathcal{O}(\log m \cdot \mathrm{MCF}(m, n) + m)$ time by using parallelization techniques as described in Section 3.1: In each of the steps 1 – 3, we need to perform $\mathcal{O}(m)$ independent comparisons[4] of linear parametric values, which can be

---

4   Recall that the number of nodes is $\mathcal{O}(m)$ due to the transformation as explained in the proof of Lemma 4.21.

done in $\mathcal{O}(\log m \cdot MCF(m,n) + m)$ time according to Lemma 3.1. Step 4 can be implemented to run in $\mathcal{O}(m)$ time according to Ahuja et al. (1993) and does not require any further callback evaluations. Hence, each of the $\mathcal{O}(m \log m)$ phases of the algorithm causes an additional overhead of $\mathcal{O}(\log m \cdot MCF(m,n) + m)$. It remains to evaluate the time needed to compute the shortest path between two nodes in the residual network. Note that the residual network can be determined in $\mathcal{O}(\log m \cdot MCF(m,n) + m)$ time as well by using the same techniques as described above in order to check if $x_e > 0$ for some $e \in E$ (since the capacities of the transformed network are infinite, each forward edge will be contained in the residual network).

Bound T_1: We are able to use a similar technique as for the steps 1 – 3 during the execution of Dijkstra's (1959) algorithm for the shortest path problem. We first apply Lemma 3.2 in order to transform the underlying multigraph of the residual network into a simple graph in $\mathcal{O}(\log\log m \log m \cdot MCF(m,n))$ time. When exploring some node v in the course of Dijkstra's (1959) algorithm, we need to evaluate the callback function once for each of the $|\delta^+(v)|$ outgoing edges $e = (v,w)$ of node v in order to update the distance label $dist(w)$ of node w to $dist(w) := \min\{dist(w), dist(v) + (a(\lambda))_e\}$. Since the underlying graph is now simple, all of the resulting comparisons are independent from each other and we only need to evaluate $\mathcal{O}(\log |\delta^+(v)|)$ minimum cost flows according to Lemma 3.1. Since Dijkstra's (1959) algorithm explores each node only once, we get the following number of minimum cost flow computations:

$$\mathcal{O}\left(\sum_{v \in V: |\delta^+(v)| > 0} (\log|\delta^+(v)| \cdot MCF(m,n) + |\delta^+(v)|)\right)$$

$$= \mathcal{O}\left(\sum_{v \in V} \log(|\delta^+(v)| + 1) \cdot MCF(m,n) + m\right)$$

$$= \mathcal{O}\left(\log\left(\prod_{v \in V}(|\delta^+(v)| + 1)\right) \cdot MCF(m,n) + m\right)$$

$$= \mathcal{O}\left(\log\left(\frac{\sum_{v \in V}(|\delta^+(v)| + 1)}{n}\right)^n \cdot MCF(m,n) + m\right)$$

$$= \mathcal{O}\left(\log\left(\frac{m+n}{n}\right)^n \cdot MCF(m,n) + m\right)$$

$$= \mathcal{O}\left(n\log\left(\frac{m}{n}\right) \cdot MCF(m,n) + m\right),$$

where the third equation follows from the inequality of arithmetic and geometric means (cf. (Cauchy, 1821)). When using the Fibonacci heap implementation of Dijkstra's (1959) algorithm due to Fredman and Tarjan (1987), we obtain

$\mathcal{O}(n \log n)$ additional minimum cost flow computation in order to maintain the Fibonacci heap. The resulting running time of

$$\mathcal{O}\left(\left(n \log \left(\frac{m}{n}\right) + n \log n + \log m \log \log m\right) \cdot MCF(m, n) + m\right)$$
$$= \mathcal{O}\left((n + \log \log m) \log m \cdot MCF(m, n) + m\right)$$

then dominates the additional overhead of $\mathcal{O}(\log m \cdot MCF(m, n) + m)$ per phase described above, which yields the claimed time bound $T_1(m, n)$.

Bound T_2: Han et al. (1992) present a parallel algorithm for the all-pairs shortest path problem on simple graphs that runs in $\mathcal{O}(\log n \log \log n)$ time on $p \in \mathcal{O}\left(\frac{f(n)}{\log n \log \log n}\right)$ processors with $f(n) = o(n^3)$. We simulate the execution of all processors sequentially until each of them either finishes its computations or halts at a comparison of linear parametric values. According to Lemma 3.1, we can resolve all comparisons simultaneously in $\mathcal{O}(\log p \cdot MCF(m, n) + p)$ time and continue the simulation. Since each processor performs at most $\mathcal{O}(\log n \log \log n)$ steps, we get a running time of

$$\mathcal{O}\left(\log n \log \log n \cdot (\log p \cdot MCF(m, n) + p)\right)$$
$$= \mathcal{O}\left(\log n \log \log n \cdot \left(\log n \cdot MCF(m, n) + \frac{f(n)}{\log n \log \log n}\right)\right)$$
$$= \mathcal{O}\left(\log^2 n \log \log n \cdot MCF(m, n) + f(n)\right).$$

In combination with the overhead of $\mathcal{O}(\log m \log \log m \cdot MCF(m, n))$ to transform the underlying multigraph into a simple graph, the claimed time bound $T_2(m, n) = \mathcal{O}((\log^2 n \log \log n + \log m \log \log m) \cdot MCF(m, n) + f(n))$ and the claim of the theorem follows. $\qquad \square$

Note that the running time of the above algorithms improves instantaneously on graph classes that allow faster strongly combinatorial algorithms for the traditional minimum cost flow problem. For example, on series-parallel graphs, a minimum cost flow can be computed in $\mathcal{O}(m \log m)$ time using an algorithm introduced by Booth and Tarjan (1992). Hence, we get the following corollary, which will turn out to be important in the subsequent chapter:

Corollary 4.23:
BCMCFP$_\mathbb{R}$ is solvable in $\mathcal{O}(m^2 \log^2 m)$ time on series-parallel graphs. $\qquad \square$

4.5 Approximability

The results of the previous sections have shown that the problem $BCMCFP_{\mathbb{R}}$ can be solved efficiently using a diverse set of approaches. In this section, we show that the problem can also be approximated efficiently. First, we will concentrate our considerations on the case of general graphs and present two FPTASs, one of which is based on a generic framework by Papadimitriou (1994) and the other is based on the generalized packing framework that was introduced in Section 3.3. The latter algorithm is then specialized to the case of acyclic graphs, on which an improved running time can be achieved.

4.5.1 General Graphs

In (Papadimitriou and Yannakakis, 2000), the authors show that an ε-approximate pareto frontier of a linear convex optimization problem with k objective functions can be determined by solving $\mathcal{O}((8Lk^2/\varepsilon)^k)$ instances of the problem with only one objective function, where L denotes the encoding-length of the largest possible objective value. Since $k = 2$ and $L \in \mathcal{O}(\log mCU) = \mathcal{O}(\log M)$ in the case of $BCMCFP_{\mathbb{R}}$, we are, thus, able to compute an ε-approximate pareto frontier in $\mathcal{O}\left(\frac{\log^2 M}{\varepsilon^2} \cdot MCF(m, n, C, U)\right)$ time. In particular, this also induces a bicriteria FPTAS for $BCMCFP_{\mathbb{R}}$ since an ε-approximate pareto frontier contains a point x_P with $b(x_P) \leqslant (1 + \varepsilon) \cdot b(x^*) \leqslant (1 + \varepsilon) \cdot B$ and $c(x_P) \leqslant \frac{1}{1+\varepsilon} \cdot c(x^*) \leqslant (1 - \varepsilon) \cdot c(x^*)$ for an optimum solution x^* of $BCMCFP_{\mathbb{R}}$. The following lemma shows that this also yields a traditional FPTAS for $BCMCFP_{\mathbb{R}}$:

Lemma 4.24:
Any bicriteria FPTAS for $BCMCFP_{\mathbb{R}}$ also induces a single-criterion FPTAS for the problem.

Proof: For each instance of $BCMCFP_{\mathbb{R}}$ with optimum solution x^*, the given bicriteria FPTAS computes a solution x with $c(x) \leqslant (1 - \varepsilon) \cdot c(x^*)$ and $b(x) \leqslant (1 + \varepsilon) \cdot b(x^*)$ in time that is polynomial in the instance size and $\frac{1}{\varepsilon}$. Since both the costs c and usage fees b are linear functions, it suffices to scale down the given solution to $x' := \frac{x}{1+\varepsilon}$. Clearly, x' is feasible since it still fulfills every flow conservation and capacity constraint and since $b(x') = \frac{1}{1+\varepsilon} \cdot b(x) \leqslant b(x^*) \leqslant B$. Moreover, for $\varepsilon' := 2\varepsilon$, it holds that

$$c(x') = \frac{1}{1+\varepsilon} \cdot c(x) \leqslant \frac{1-\varepsilon}{1+\varepsilon} \cdot c(x^*) = \frac{1-\frac{\varepsilon'}{2}}{1+\frac{\varepsilon'}{2}} \cdot c(x^*)$$

$$= \frac{(1-\frac{\varepsilon'}{2}) \cdot (1-\frac{\varepsilon'}{2})}{(1+\frac{\varepsilon'}{2}) \cdot (1-\frac{\varepsilon'}{2})} \cdot c(x^*) = \frac{1-\varepsilon' + \frac{(\varepsilon')^2}{4}}{1 - \frac{(\varepsilon')^2}{4}} \cdot c(x^*)$$

$$\leqslant (1-\varepsilon') \cdot c(x^*),$$

where the last inequality follows from the fact that $c(x^*) < 0$. $\qquad\square$

The result of Lemma 4.24 in combination with the above discussion yields the following corollary:

Corollary 4.25:
There is an FPTAS for BCMCFP$_\mathbb{R}$ running in $\mathcal{O}\left(\frac{\log^2 M}{\varepsilon^2} \cdot \text{MCF}(m, n, C, U)\right)$ time. $\qquad\square$

The above FPTAS has a satisfactory theoretical running time and can also be used to derive an ε-approximate frontier for BCMCFP$_\mathbb{R}$. However, the running time is only weakly polynomial and is connected with large coefficients that are hidden by the $\mathcal{O}$-notation. We now show how we can obtain a tailored FPTAS for BCMCFP$_\mathbb{R}$ with a *strongly* polynomial running time. In order to do so, we make use of the generalized packing framework that was introduced in Section 3.3.

Theorem 4.26:
There is an FPTAS for BCMCFP$_\mathbb{R}$ running in $\widetilde{\mathcal{O}}\left(\frac{1}{\varepsilon^2} \cdot (nm^2 + n^3m)\right)$ time.

Proof: Consider an equivalent circulation based variant of BCMCFP$_\mathbb{R}$ that can be obtained by inserting an edge with infinite capacity, zero costs, and zero usage fees between t and s. Since each circulation in this transformed instance is a traditional network circulation and since $b(C) > 0$ for each cycle C, each optimal circulation can be decomposed into at most m circulations on negative cycles according to the flow decomposition theorem for traditional flows (cf. (Ahuja et al., 1993)). Hence, if S is the set of all flows $x^{(l)}$ with unit flow value on negative cost cycles, we have that each solution x lies in the cone that is generated by the flows in S.

As it was already noted in Example 3.12, we are able to make use of Theorem 3.10 in order to obtain a fully polynomial-time approximation scheme for BCMCFP$_\mathbb{R}$ with an overall running time of $\mathcal{O}\left(\frac{1}{\varepsilon^2} \cdot m \log m \cdot (nm)^2\right)$. This running time can be significantly improved by the following observation: As it was shown in Section 3.3.2, the sign oracle is incorporated into Megiddo's parametric search in order to determine a minimizer of

$$\min_{\substack{l \in \{1, \dots, k\} \\ c^T x^{(l)} > 0}} \frac{a^T x^{(l)}}{c^T x^{(l)}}. \tag{3.6}$$

for a positive cost vector a and a vector c (cf. page 27). In the case of $BCMCFP_{\mathbb{R}}$, this reduces to the determination of a minimum ratio cycle C. Megiddo (1983) derived an algorithm that determines a minimum ratio cycle in a *simple* graph in $\mathcal{O}(n^3 \log n + nm \log^2 n \log \log n)$ time by making use of a parallel algorithm for the all-pair shortest path problem in combination with Karp's minimum mean cycle algorithm (Karp, 1978) as a negative cycle detector in his parametric search. In order to comply with our setting of multigraphs, we first need to apply Lemma 3.2 with the minimum mean cycle algorithm as a callback to the underlying graph, which yields a running time of $\mathcal{O}(nm \log m \log \log m)$. In total, we get that

$$A(m, n) = \mathcal{O}(nm \log m \log \log m + n^3 \log n + nm \log^2 n \log \log n).$$

Thus, incorporated in the general packing framework, we obtain an overall running time of

$$\mathcal{O}\left(\frac{1}{\varepsilon^2} \cdot m \log m \cdot A(m, n)\right) = \tilde{\mathcal{O}}\left(\frac{1}{\varepsilon^2} \cdot (nm^2 + n^3 m)\right). \qquad \square$$

4.5.2 Acyclic Graphs

We now show how we can improve the running time of the FPTAS described in Theorem 4.26 for the case of an acyclic graph G. Since there are no cycles in G, we only need to repeatedly determine *minimum ratio s-t-paths* rather than minimum ratio cycles. This, however, can be done more efficiently, as shown in the following lemma:

Lemma 4.27:
Let $d^{(1)}: E \to \mathbb{R}$ and $d^{(2)}: E \to \mathbb{R}$ be two cost functions with $d_e^{(1)} := d^{(1)}(e)$ and $d_e^{(2)} := d^{(2)}(e)$ for each $e \in E$ and assume that $\sum_{e \in P} d_e^{(2)} > 0$ for each s-t-path P. An s-t-path P^* that minimizes the ratio $\frac{\sum_{e \in P} d_e^{(1)}}{\sum_{e \in P} d_e^{(2)}}$ among all s-t-paths P can be determined in $\mathcal{O}(m \log m \log \log m + nm \log n)$ time on acyclic graphs.

Proof: Let $\mathcal{P}$ denote the set of all s-t-paths in the underlying graph G. As it was shown in Lemma 3.8, we can restrict our considerations to the problem $\min_{P \in \mathcal{P}} \sum_{e \in P} d_e^{(\lambda)}$ with $d_e^{(\lambda)} := d_e^{(1)} - \lambda \cdot d_e^{(2)}$ for each $e \in E$ and some parameter λ: For some given value of the parameter λ, it holds that $\min_{P \in \mathcal{P}} \sum_{e \in P} d_e^{(\lambda)}$ is negative (positive) if and only if the value of λ is smaller (larger) than the value λ^* that leads to an optimum solution P^* to $\min_{P \in \mathcal{P}} \frac{\sum_{e \in P} d_e^{(1)}}{\sum_{e \in P} d_e^{(2)}}$. Hence, by simulating the shortest path algorithm for acyclic graphs with edge lengths $d^{(\lambda)}$ for a symbolic value of λ using Megiddo's parametric search technique, it is possible to determine the optimum solution P^* in $\mathcal{O}(m^2)$ time.

We can improve this running time by first applying Lemma 3.2 to the underlying multigraph in order to obtain a simple graph with the same shortest paths as in G in $\mathcal{O}(m \log m \log \log m)$ time. In this simple graph, we simulate the shortest path algorithm for acyclic graphs, which initially sets the distance label dist(v) of each node v to infinity. The algorithm then investigates the nodes in the order of a topological sorting and, for each outgoing edge $e = (v, w)$ of some node $v \in V \setminus \{t\}$ in this sorting, updates the distance label dist(w) of node w to $\min\{\text{dist}(w), \text{dist}(v) + l_e\}$ where l_e denotes the length of edge e. Similar to the proof of Theorem 4.22, since the edges in $\delta^+(v)$ head to different nodes in the simple graph, all of these comparisons are independent from each other and can be simultaneously resolved in $\mathcal{O}(\log |\delta^+(v)| \cdot \text{SP}(m, n) + |\delta^+(v)|)$ time according to Lemma 3.1. This results in a running time for the parametric shortest path computation of

$$\mathcal{O}\left(\sum_{v \in V \setminus \{t\}} (\log |\delta^+(v)| \cdot \text{SP}(m, n) + |\delta^+(v)|)\right) = \mathcal{O}(n \log n \cdot \text{SP}(m, n) + m)$$

$$= \mathcal{O}(nm \log n),$$

which in combination with the overhead of $\mathcal{O}(m \log m \log \log m)$ for the transformation into a simple graph shows the claim. $\qquad \square$

By incorporating the results of Lemma 4.27 into the generalized packing framework as introduced in Section 3.3, we immediately get the following corollary:

Corollary 4.28:
There is an FPTAS for BCMCFP$_\mathbb{R}$ that runs in $\mathcal{O}(\frac{1}{\varepsilon^2} \cdot m^2 \log m \cdot (\log m \log \log m + n \log n))$ time on acyclic graphs. $\qquad \square$

4.6 Conclusion

We studied a natural extension of the minimum cost flow problem by a budget constraint that restricts the usage of edges based on a second kind of costs. It was shown that the problem has a wide variety of possible applications, although only a little amount of literature is published at present. As it follows directly by Vaidya (1989), the problem can be solved in weakly polynomial time by the use of interior point methods. Nevertheless, since we are interested in efficient *combinatorial* algorithms, we presented three such algorithms that exploit the discrete structure of the underlying problem.

On the one hand, in Section 4.3, we developed a fully combinatorial network simplex algorithm and showed that it can be implemented to run in pseudo-polynomial time. In doing so, we derived optimality criteria for the problem and provided rules for the choice of both the entering and leaving edge in order to improve the running time and to avoid cycling, respectively.

On the other hand, in Section 4.4, we presented an interpretation of the problem as a bicriteria minimum cost flow problem, which enabled us to reduce the problem $BCMCFP_R$ to the computation of a series of traditional minimum cost flows. In particular, we obtained a weakly polynomial-time algorithm that computes $\mathcal{O}(\log M)$ minimum cost flows and a strongly polynomial-time algorithm based on Megiddo's parametric search techniques using $\widetilde{\mathcal{O}}(nm)$ minimum cost flow computations.

Finally, we studied the approximability of the problem in Section 4.5. As it was shown, we can approximate the problem in weakly polynomial time by making use of the generic approximation framework due to Papadimitriou and Yannakakis (2000). Furthermore, we were able to provide an FPTAS with a strongly polynomial time bound based on the generalized packing framework as introduced in Section 3.3. This algorithm was subsequently refined for the case of acyclic graphs in order to obtain an even faster algorithm. An overview of the results of this chapter in given in Table 4.1.

The introduced model raises several questions for future research. In particular, it is worth investigating the empirical performance of the presented network simplex algorithm. Using similar ideas as in (Ahuja et al., 2002; Orlin, 1997; Tarjan, 1997), it may be possible to improve the overall running time of the procedure. Moreover, as in the case of the network simplex algorithm, one may try to apply other combinatorial algorithms for the traditional minimum cost flow problem to the case of $BCMCFP_R$. In particular, a variant of the successive shortest path algorithm for the traditional minimum cost flow problem (cf. (Ahuja et al., 1993)) in which one packs minimum ratio paths rather than shortest path seems promising.

General Graphs	Acyclic Graphs	Series-Parallel Graphs
Theorem 4.4: Solvable in $\mathcal{O}(m^{2.5} \cdot \log M)$ time (interior point)	$\longrightarrow$	$\longrightarrow$
Theorem 4.17: Solvable in $\mathcal{O}(n^4 m^2 CUB^2 \cdot \log(mCUB))$ time (network simplex)	$\longrightarrow$	$\longrightarrow$
Theorem 4.20: Solvable in $\mathcal{O}(\log M \cdot MCF(m,n,C,U))$ time (bicriteria approach)	$\longrightarrow$	$\longrightarrow$
Theorem 4.22: Solvable in $\tilde{\mathcal{O}}(\min\{nm \cdot MCF(m,n), m \cdot MCF(m,n) + m \cdot f(n)\})$ time (bicriteria approach)	$\longrightarrow$	**Corollary 4.23:** Solvable in $\mathcal{O}(m^2 \log^2 m)$ time (bicriteria approach)
Corollary 4.25: FPTAS in $\mathcal{O}\left(\frac{\log^2 M}{\varepsilon^2} \cdot MCF(m,n,C,U)\right)$ t (approximate pareto frontier approach)	$\longrightarrow$	$\longrightarrow$
Theorem 4.26: FPTAS in $\tilde{\mathcal{O}}\left(\frac{1}{\varepsilon^2} \cdot (nm^2 + n^3 m)\right)$ time (generalized packing framework)	**Corollary 4.28:** FPTAS in $\mathcal{O}(\frac{1}{\varepsilon^2} \cdot m^2 \log m \cdot (\log m \log \log m + n \log n))$ time (generalized packing framework)	$\longrightarrow$

Table 4.1: The summarized results for the continuous budget-constrained minimum cost flow problem in Chapter 4. Implied results are denoted with gray arrows.

5 | Budget-Constrained Minimum Cost Flows: The Discrete Case

We now consider two discrete variants of the budget-constrained minimum cost flow problem that was investigated in the previous chapter. We show that both variants may be interpreted as network improvement problems, which yields several fields of applications. As a first variant, we prove that the problem becomes both weakly $\mathcal{NP}$-hard to solve and approximate if the usage fees are induced in integral units. For the case of series-parallel graphs, we derive a pseudo-polynomial-time exact algorithm. Moreover, we present an interesting interpretation of the problem on extension-parallel graphs as a knapsack problem and provide both an approximation algorithm and (fully) polynomial-time approximation schemes. Finally, as a second discrete variant, we investigate a binary case in which a fee is incurred for a positive flow on an edge, independently of the flow's magnitude. For this case, the problem becomes strongly $\mathcal{NP}$-hard to solve, but still solvable in pseudo-polynomial-time on series-parallel graphs and easy to approximate under several restrictions on extension-parallel graphs.

This chapter is based on joint work with Sven O. Krumke and Clemens Thielen (Holzhauser et al., 2016a).

5.1 Introduction

As it was shown in the previous chapter, the budget-constrained minimum cost flow problem embodies a natural extension of the traditional minimum cost flow problem that allows to model a large variety of real-world problems. However, one can think of scenarios in which the continuous connection between the flow and the induced usage fee on an edge that was investigated in Chapter 4 may be insufficient, for example in order to model discrete levels of usage fees or fixed installation costs that are incurred independently from the actual amount of flow.

In this chapter, we investigate two discrete variants of the budget-constrained minimum cost flow problem. In the first variant, the usage fee on each edge is charged in integral steps based on the flow value on the edge that is rounded up to the next integer value. In the second variant, the usage fees on the edges are fixed for positive flow values, independently from the actual magnitude of the flow. Using these assessments, the budget-constrained minimum cost flow problem can be interpreted as a network improvement problem in which the edges need first to be upgraded to a suitable positive value before a minimum cost flow can travel through the network. These edge upgrades then induce costs that are bounded by a given budget.

The investigated discrete variants of the budget-constrained minimum cost flow problem are motivated by the fact that the possibility to upgrade edges subject to a maximum budget is ubiquitous in many applications of the traditional minimum cost flow problem such as transportation problems. Moreover, as it was noted in the previous chapter, it is well-known that the *maximum dynamic flow problem* can be formulated as a minimum cost flow problem, so our model contains the budget-constrained maximum dynamic flow problem as a special case. Among others, this problem has applications in the planning of (discrete) pipe diameters in urban drainage systems, where pipes need to be built or upgraded in order to guarantee that a sufficient amount of wastewater can be transported to sewage plants in a given time horizon without violating the budget of the city (cf. (Spellman, 2013)). Moreover, as the shortest path problem is a special case of the traditional minimum cost flow problem, the discrete variant of the budget-constrained minimum cost flow problem contains the well-known *constrained shortest path problem* as a special case when sending one (integral) unit of flow from one node to another (cf. (Garey and Johnson, 1979; Ziegelmann, 2001)).

A large part of this chapter will set the focus on the variants of the problem on extension-parallel graphs. Note that the urban drainage systems mentioned above are usually formed as in-trees leading to the sewage plant (cf. (Spellman, 2013)), which are basically extension-parallel graphs if the leaves of the corresponding trees are connected with the source using artificial edges. Hence, our algorithms may in particular be used to compute exact and efficient approximate solutions to this problem.

5.1.1 Previous Work

Krumke and Schwarz (1998) study the problem of finding a maximum flow in the case that the capacity of each edge can be improved using a given budget. To do so, they differentiate between the same three variants of how to calculate the upgrade costs that we use in this and the previous chapter. The authors show that the continuous and the integral variant of the problem are easy to solve on general graphs while the binary variant becomes strongly $\mathcal{NP}$-complete to solve. The same differentiation was later used by Krumke et al. (1999) who investigate the problem of finding a minimum cost improvement strategy of the edge capacities in order to allow a flow of a given value to travel through the network. The authors show that this problem is strongly $\mathcal{NP}$-complete to solve on general graphs and easy to approximate on series-parallel graphs for the case that the edges may only be upgraded to their maximum upgrade value, which corresponds to our binary variant of the problem.

Problems related to the binary case of the budget-constrained minimum cost flow problem were furthermore investigated by Hochbaum and Segev (1989) and recently by Duque et al. (2013). In the first paper, the authors introduce a model in which the usage of each arc induces not only costs that depend linearly on the flow but also fixed costs. A similar problem, in which a maximum flow is computed while fixed costs are incurred by the usage of edges that must fulfill a given budget constraint, was introduced in (Garey and Johnson, 1979, Problem ND32). In (Duque et al., 2013), the authors show that the problem of finding a minimum cost flow in case that the capacity of each edge can be upgraded to one of several discrete levels using a given budget is $\mathcal{NP}$-complete to solve and provide heuristics along with computational results.

5.1.2 Chapter Outline

After a formal definition of the two discrete variants of the budget-constrained minimum cost flow problem in Section 5.2, we concentrate on the integral case in Section 5.3 and show that the problem is weakly $\mathcal{NP}$-complete to solve, even on extension-parallel graphs and $\mathcal{NP}$-hard to approximate, even on series-parallel graphs. Moreover, we present a pseudo-polynomial-time exact algorithm that solves the problem on series-parallel graphs. In Section 5.4, we show that the problem on extension-parallel graphs has an interpretation as a novel extension of the bounded knapsack problem (Kellerer et al., 2004). Using this fact, we are able to adapt several results that are valid for the traditional bounded knapsack problem. In particular, we present a 2-approximation algorithm, which is then incorporated into a PTAS and, under the restriction of polynomially bounded capacities, into an FPTAS. Moreover, we show that an approximate pareto frontier can be determined efficiently and that we can use this result in order to derive a bicriteria FPTAS. We also identify a special case of the problem that is solvable in polynomial time. Finally, in Section 5.5, we show that the binary variant becomes strongly $\mathcal{NP}$-complete to solve and weakly $\mathcal{NP}$-hard to approximate. However, as in the integral case, we are able to derive a pseudo-polynomial-time exact algorithm for the problem on series-parallel graphs and an FPTAS for the problem on extension-parallel graphs if the edge capacities are polynomially bounded. An overview of the results of this chapter is given in Table 5.1 and Table 5.2 on page 114.

5.2 Preliminaries

Recall that in the continuous case that was investigated in Chapter 4, the total usage fee on an edge e that transports x_e units of flow was given by $b_e \cdot \dot{x}_e$. In the following, we distinguish between two discrete variants of how the overall fee $b(x)$ of a flow x is calculated. On the one hand, in the *integral case*, the fees are calculated as in the case of BCMCFP$_\mathbb{R}$ but the amount of flow is rounded up to the next integer, i.e., the overall usage fee is given by $b_\mathbb{N}(x) := \sum_{e \in E} b_e \cdot \lceil x_e \rceil$. On the other hand, in the *binary case*, the fee is incurred on each edge as long as the flow on the edge is positive, i.e., $b_\mathbb{B}(x) := \sum_{e \in E} b_e \cdot u_e \cdot \mathrm{sgn}(x_e)$. These two variants give rise to the following two definitions of the (discrete) budget-constrained minimum cost flow problem:

Definition 5.1 (Budget-constrained minimum cost flow problem (BCMCFP$_\mathbb{N}$, BCMCFP$_\mathbb{B}$)):

INSTANCE: Directed graph $G = (V, E)$ with source $s \in V$, sink $t \in V$, capacities $u_e \in \mathbb{N}$, costs $c_e \in \mathbb{Z}$, and usage fees $b_e \in \mathbb{N}_{\geqslant 0}$ on the edges $e \in E$ and a budget $B \in \mathbb{N}_{\geqslant 0}$.

TASK:

BCMCFP$_\mathbb{N}$: Determine a budget-constrained minimum cost flow for $b := b_\mathbb{N}$.

BCMCFP$_\mathbb{B}$: Determine a budget-constrained minimum cost flow for $b := b_\mathbb{B}$.

$\triangleleft$

As in the previous chapter, we use the definitions $C := \max_{e \in E} |c_e|$, $U := \max_{e \in E} u_e$, and $B := \max_{e \in E} b_e$ when making statements about time complexities in the following.

A more intuitive interpretation of these two discrete variants of budget-constrained minimum cost flows can be obtained by considering the budget-constrained minimum cost flow problem as a network improvement problem in the following way: Initially, each edge $e \in E$ has a capacity of zero and a maximum *upgrade capacity* of u_e. In order to send some specific amount of flow x_e over e, it is necessary to upgrade its capacity up to a sufficient amount $y_e \geqslant x_e$, which generates *upgrade costs* $b_e \cdot y_e$. The upgrades y_e will then be integral for BCMCFP$_\mathbb{N}$ while the variant BCMCFP$_\mathbb{B}$ can then be seen as an "all-or-nothing" version in which each edge must either remain at zero capacity or be upgraded to its maximum capacity. The problems BCMCFP$_\mathbb{N}$ and BCMCFP$_\mathbb{B}$ then reduce to the determination of an *upgrade profile*, i.e., a function $y \colon E \to \mathbb{N}_{\geqslant 0}$ with $0 \leqslant y_e \leqslant u_e$ and $\sum_{e \in E} b_e \cdot y_e \leqslant B$ for $y_e := y(e)$. Among all such upgrade profiles, the aim is to determine an upgrade profile for which a traditional minimum cost flow in the induced graph G_y with edge capacity y_e for each $e \in E$ attains its minimum value. Since, without loss of generality, traditional minimum cost flows are integral for integral edge capacities (cf. (Ahuja et al., 1993)), this observation yields the following corollary:

Corollary 5.2:
Without loss of generality, each optimal flow is integral in the case of $\text{BCMCFP}_\mathbb{N}$ and $\text{BCMCFP}_\mathbb{B}$. □

Based on the above discussion, we will use the term *upgrade costs* instead of *usage fees* for the values b_e for $e \in E$ throughout this chapter. Using the definitions from above, we can formulate the discrete budget-constrained minimum cost flow problem as an integer linear program as follows:

$$\min \sum_{e \in E} c_e \cdot x_e \tag{5.1a}$$

$$\text{s.t.} \sum_{e \in \delta^-(v)} x_e - \sum_{e \in \delta^+(v)} x_e = 0 \qquad \text{for all } v \in V \setminus \{s, t\}, \tag{5.1b}$$

$$\sum_{e \in E} y_e \cdot b_e \leqslant B, \tag{5.1c}$$

$$0 \leqslant x_e \leqslant y_e \qquad \text{for all } e \in E, \tag{5.1d}$$

$$0 \leqslant y_e \leqslant u_e \qquad \text{for all } e \in E. \tag{5.1e}$$

The variables y_e for $e \in E$ can be seen as the upgrades that are necessary to allow the flow x to pass through the network. In the case of $\text{BCMCFP}_\mathbb{N}$, we additionally restrict these amounts to be integral:

$$y_e \in \mathbb{N}_{\geqslant 0} \qquad \text{for all } e \in E. \tag{5.1f}$$

Accordingly, the upgrade amounts y_e are forced to be either zero or u_e in the case of $\text{BCMCFP}_\mathbb{B}$:

$$y_e \in \{0, u_e\} \qquad \text{for all } e \in E. \tag{5.1g}$$

Although this linear integer program may be solved using standard techniques such as branch-and-cut methods (Wolsey and Nemhauser, 2014), we will develop both combinatorial exact and approximate solutions to the problems in the following. As in the previous chapter, note that we explicitly allow negative costs but do not specify a flow value, which does not impose any restrictions (cf. Section 4.2). As a consequence, the problem always has a non-positive optimal solution.

5.3 Integral Case

We begin our considerations with the case of integral capacity upgrades, i.e., we assume that $b(x) = b_\mathbb{N}(x) = \sum_{e \in E} b_e \cdot \lceil x_e \rceil$ for each flow x. As we show in the following theorems, the problem turns out to be weakly $\mathcal{NP}$-complete to solve and $\mathcal{NP}$-hard to approximate in this setting. Nevertheless, we are able to provide a pseudo-polynomial-time exact algorithm for the problem on series-parallel graphs.

5.3.1 Complexity

We begin with a statement about the complexity of $\mathrm{BCMCFP}_\mathbb{N}$. In contrast to the problem $\mathrm{BCMCFP}_\mathbb{R}$ that was considered in Chapter 4, the problem becomes weakly $\mathcal{NP}$-complete to solve in the case of integral upgrades:

Theorem 5.3:
$\mathrm{BCMCFP}_\mathbb{N}$ is weakly $\mathcal{NP}$-complete to solve, even when $u_e = 1$ for each $e \in E$ and the graph consists of parallel edges only.

Proof: $\mathrm{BCMCFP}_\mathbb{N}$ is obviously contained in $\mathcal{NP}$ since we can easily check if a given flow x (which has a polynomially bounded encoding length) is feasible and has costs less or equal a given bound in polynomial time. We show the $\mathcal{NP}$-hardness by using a reduction from the weakly $\mathcal{NP}$-complete problem SUBSETSUM, which can be defined as follows (Garey and Johnson, 1979, Problem SP13):

INSTANCE: Finite set $\{a_1, \ldots, a_k\}$ of k positive integers and a positive integer A.

QUESTION: Is there a subset $I \subseteq \{1, \ldots, k\}$ such that $\sum_{i \in I} a_i = A$?

Given an instance of SUBSETSUM, we construct an instance of $\mathrm{BCMCFP}_\mathbb{N}$ as follows: For each $i \in \{1, \ldots, k\}$, we introduce an edge e_i from the source s to the sink t with cost $c_{e_i} := -a_i$, upgrade cost $b_{e_i} := a_i$, and capacity $u_{e_i} := 1$. The budget is set to $B := A$.

We now show that there exists a feasible integral flow x with $c(x) \leqslant -A$ if and only if the given instance of SUBSETSUM is a YES-instance.

Let x be a feasible integral flow in the constructed network with $c(x) \leqslant -A$. Let $I := \{i \in \{1, \ldots, k\} : x_{e_i} = 1\}$ be the index set of edges carrying one unit of flow. Since $b(x) \leqslant B$, we have

$$\sum_{i \in I} a_i = \sum_{i=1}^{k} b_{e_i} \cdot x_{e_i} = \sum_{e \in E} b_e \cdot x_e = b(x) \leqslant B = A.$$

Similarly, since $c(x) \leqslant -A$, we obtain

$$\sum_{i \in I} a_i = \sum_{i=1}^{k} -c_{e_i} \cdot x_{e_i} = \sum_{e \in E} -c_e \cdot x_e = -c(x) \geqslant A.$$

Thus, we get that $\sum_{i \in I} a_i = A$, so I is a solution of the given instance of SUBSETSUM.

On the other hand, for a solution I of the given instance of SUBSETSUM, we get a feasible integral flow as follows: We set $x_{e_i} := 1$ for $i \in I$ and $x_{e_j} := 0$ for $j \notin I$. The upgrade costs are then given by $\sum_{i \in I} b_{e_i} = \sum_{i \in I} a_i = A = B$ and the costs evaluate to $c(x) = \sum_{i \in I} c_{e_i} = \sum_{i \in I} -a_i = -A$, which proves the claim. □

5.3.2 Approximability

Since the problem BCMCFP_N is $\mathcal{NP}$-complete to solve, even on extension-parallel graphs according to Theorem 5.3, one might still hope for the existence of efficient approximation algorithms. However, as it will be shown in the following theorem, the problem is even $\mathcal{NP}$-hard to approximate on series-parallel graphs:

Theorem 5.4:
BCMCFP_N is weakly $\mathcal{NP}$-hard (to solve and) to approximate within constant factors, even on bipartite series-parallel graphs and when $u_e = 1$ for each $e \in E$.

Proof: We show the claim by using a reduction from the problem EVENODDPARTITION, which is known to be weakly $\mathcal{NP}$-complete (Garey and Johnson, 1979, Problem SP12):

INSTANCE: Finite set $\{a_1, \ldots, a_{2k}\}$ of positive integers.
QUESTION: Is there a subset $I \subseteq \{1, \ldots, 2k\}$ such that $\sum_{i \in I} a_i = \sum_{i \notin I} a_i$ and $|I \cap \{a_{2j-1}, a_{2j}\}| = 1$ for each $j \in \{1, \ldots, k\}$?

We now show that an instance of EVENODDPARTITION can be transformed to an equivalent instance of BCMCFP_N in polynomial time. Afterwards, we show that any approximation algorithm for BCMCFP_N, when applied to the resulting instance of BCMCFP_N, can be used to decide whether the original instance of EVENODD-PARTITION is a YES-instance, which proves the claim.

Given an instance of EVENODDPARTITION, we construct an instance of BCMCFP_N as follows: For $j \in \{1, \ldots, k+1\}$, we introduce a node v_j. Between each pair of nodes v_j and v_{j+1}, we introduce two parallel edges e_{2j-1} and e_{2j}. The former edge has costs $c_{e_{2j-1}} := a_{2j-1}$ and upgrade costs $b_{e_{2j-1}} := a_{2j}$ while the latter edge has costs $c_{e_{2j}} := a_{2j}$ and upgrade costs $b_{e_{2j}} := a_{2j-1}$. Each of these edges e_i, $i \in \{1, \ldots, 2k\}$, has a capacity of $u_{e_i} := 1$. Furthermore, we insert an edge e_0 between v_{k+1} and the sink t with capacity $u_{e_0} := 1$, upgrade costs $b_{e_0} := 0$ and costs $c_{e_0} := -A - 1$ where $A := \frac{1}{2} \cdot \sum_{i=1}^{2k} a_i$ is the size of each of the two partitions in a solution of EVENODDPARTITION. Moreover, we identify the source s with the node v_1 and set the budget to $B := A$. The constructed network is shown in Figure 5.1.

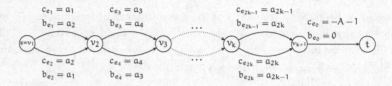

Figure 5.1: The constructed instance for the reduction of EVENODDPARTITION to BCMCFP$_N$.

We claim that there exists a flow x with costs $c(x) = -1$ if the given instance of EVEN-ODDPARTITION is a YES-instance and that $c(x) = 0$ for each feasible flow x else. Before proving this result, note that the maximum flow value of a flow x in the constructed network is at most one because of the edge with capacity one that leads to the sink and since the budget allows only one v_1-t-path to be upgraded.

First assume that the given instance of EVENODDPARTITION is a YES-instance, i.e., there is a set $I \subseteq \{1, \dots, 2k\}$ such that $\sum_{i \in I} a_i = A = \sum_{i \notin I} a_i$ and $|I \cap \{a_{2j-1}, a_{2j}\}| = 1$ for each $j \in \{1, \dots, k\}$. By sending one unit of flow through the edges e_i for $i \in I$ and the edge e_0, we get an integral flow x that satisfies

$$c(x) = \sum_{e \in E} c_e \cdot x_e = \sum_{i \in I} c_{e_i} \cdot x_{e_i} + c_{e_0} \cdot x_{e_0} = \sum_{i \in I} a_i - A - 1 = -1$$

and, since the costs and upgrade costs of two parallel edges are contrarily,

$$b(x) = \sum_{e \in E} b_e \cdot x_e = \sum_{i \in I} b_{e_i} \cdot x_{e_i} + b_{e_0} \cdot x_{e_0} = \sum_{i \notin I} a_i + 0 = A = B.$$

Hence, the budget is not exceeded and the costs fulfill $c(x) = -1$ as claimed.

Assume conversely that the given instance of EVENODDPARTITION is a No-instance, i.e., $\sum_{i \in I} a_i \neq A$ for each $I \subseteq \{1, \dots, 2k\}$ with $|I \cap \{a_{2j-1}, a_{2j}\}| = 1$ for $j \in \{1, \dots, k\}$. We show that every feasible flow x has cost $c(x) = 0$. Since each such set I can be identified with a path P from v_1 to v_{k+1}, we get that either the costs or the upgrade costs per unit of flow on each such path P are greater or equal to $A + 1$. Since we must satisfy the budget constraint, we assume that flow is sent on a v_1-v_{k+1}-path with upgrade costs of at most $A - 1$ and costs of at least $A + 1$. But then the total cost per unit of flow on P and e_0 is at least $(A + 1) + (-A - 1) = 0$, i.e., every feasible flow x has cost $c(x) = 0$ in this case (since the zero-flow is allowed).

Thus, the costs of a budget-constrained minimum cost flow are at most -1 if the given instance of EVENODDPARTITION is a YES-instance and zero else.

Now consider an approximation algorithm that computes a feasible flow x in the constructed network with costs $c(x)$ fulfilling

$$c(x^*) \leqslant c(x) \leqslant \frac{1}{\alpha} \cdot c(x^*),$$

where x^* is the optimal flow and $\alpha \in [1, \infty)$ is the performance guarantee. If $c(x) < 0$, it must hold that $c(x^*) \leqslant -1$ due to the left inequality. On the other hand, if $c(x) = 0$, we get that $c(x^*) = 0$ because of the right inequality. Hence, we can decide if the underlying instance of EVENODDPARTITION is a YES-instance by checking the sign of $c(x)$, which shows the $\mathcal{NP}$-hardness of approximation. $\square$

Note that the proofs of Theorem 5.3 and Theorem 5.4 both use parallel edges in the reductions. However, both proofs would work analogously after inserting an artificial node in the middle of each parallel edge.

5.3.3 Exact Algorithm

The results in the previous two subsections have shown that the problem BCMCFP_N is *at least* weakly $\mathcal{NP}$-complete to solve on series-parallel graphs. In this subsection, we show that BCMCFP_N is solvable in pseudo-polynomial time on series-parallel graphs. In particular, as it is well known, this shows that the problem is not strongly $\mathcal{NP}$-complete to solve (unless $\mathcal{P} = \mathcal{NP}$) such that the complexity of the problem on series-parallel graphs is bounded both from above and below.

The proof of the following theorem uses a result by Bein et al. (1985), who showed that a traditional minimum cost flow on a series-parallel graph can be constructed in a greedy manner by repeatedly filling shortest paths with positive remaining capacity (without considering backward edges in a residual network). In the following, we refer to such a flow as being *positive only on shortest paths*.

Theorem 5.5:
BCMCFP_N is solvable in pseudo-polynomial time $\mathcal{O}(nmCUB^2 \cdot (n^2C + m^2U))$ on series-parallel graphs.

Proof: We first show that the costs of a budget-constrained minimum cost flow can be computed in pseudo-polynomial time before showing how the actual flow can be computed within the same running time. Therefore, let $A_{G'}(c, b, f)$ denote the minimum cost of a flow in a series-parallel graph G' with flow value exactly f, upgrade cost at most b, and that has a flow decomposition into s-t-paths P with cost of $c(P) := \sum_{e \in P} c_e \leqslant c$ per unit of flow. If no such flow exists, we set $A_{G'}(c, b, f) := +\infty$.

Note that, for a given integral upgrade profile y, the result of Bein et al. (1985) implies that there always exists an optimal solution x^* of the traditional minimum cost flow problem in the graph G_y such that x^* uses only shortest paths in G_y. Hence, whenever we consider a fixed upgrade profile y for some series-parallel subgraph G', we may always assume that the corresponding optimal flow in G' uses only paths that are shortest paths in $(G')_y$.

The proof of the theorem works by a recursive computation of the values $A_{G'}(c, b, f)$ that is based on a bottom-up traversal of a fixed decomposition tree T of the given series-parallel graph G. According to the definition of series-parallel graphs (cf. Section 2.3), we differentiate between the following three types of series-parallel subgraphs G' that correspond to nodes in the decomposition tree of G:

Single edge $G' = e$:

If the graph G' only consists of a single edge e and the cost c_e exceeds the given bound c, the only feasible flow is the zero flow. Thus, we have $A_{G'}(c, b, f) = 0$ if $f = 0$ and $A_{G'}(c, b, f) = +\infty$ else in this case. Otherwise, the flow value is bounded by the capacity u_e and the maximum capacity upgrade $\left\lfloor \frac{b}{b_e} \right\rfloor$ that does not exceed the budget b.[1] The demanded flow value can be achieved if and only if this bound is at least f. Thus, the value $A_{G'}(c, b, f)$ is given by

$$A_{G'}(c, b, f) := \begin{cases} f \cdot c_e, & \text{if } \left(c_e \leqslant c \wedge f \leqslant u_e \wedge f \leqslant \left\lfloor \frac{b}{b_e} \right\rfloor \right) \vee f = 0, \\ +\infty, & \text{else.} \end{cases}$$

Series composition $G' = G_1 \circ G_2$:

We show that the value $A_{G'}(c, b, f)$ in a graph G' that is the series composition of two series-parallel graphs $G_1 = (V_1, E_1)$ and $G_2 = (V_2, E_2)$ is given by

$$A_{G'}(c, b, f) := \min_{\substack{0 \leqslant b_1 \leqslant b \\ \max\{c,0\} - nC \leqslant c_1 \leqslant \min\{c,0\} + nC}} A_{G_1}(c_1, b_1, f) + A_{G_2}(c - c_1, b - b_1, f). \quad (5.2)$$

"$\leqslant$": Consider some fixed values for b_1 and c_1. If one of the terms on the right-hand side of equation (5.2) is infinite, the claim clearly follows. Otherwise, both $A_{G_1}(c_1, b_1, f) < +\infty$ and $A_{G_2}(c - c_1, b - b_1, f) < +\infty$, so there exist feasible flows $x^{(1)}$ in G_1 and $x^{(2)}$ in G_2 of flow value f that satisfy the respective bounds

---

1   We assume that $\left\lfloor \frac{b}{b_e} \right\rfloor = +\infty$ if $b_e = 0$.

on the cost and upgrade cost and have total cost $A_{G_1}(c_1, b_1, f)$ and $A_{G_2}(c - c_1, b - b_1, f)$, respectively. Hence, setting $x_e := x_e^{(1)}$ for $e \in E_1$ and $x_e := x_e^{(2)}$ for $e \in E_2$ yields a feasible flow x in G' of flow value f and upgrade cost at most $b_1 + (b - b_1) = b$ on paths with cost at most $c_1 + (c - c_1) = c$ per unit of flow, which shows that $A_{G'}(c, b, f) \leqslant c(x) = A_{G_1}(c_1, b_1, f) + A_{G_2}(c - c_1, b - b_1, f)$.

"$\geqslant$": If $A_{G'}(c, b, f) = +\infty$, the claim again follows. Otherwise, consider a budget-constrained minimum cost flow x in G' with upgrade cost $b(x) \leqslant B$ and flow value $\mathrm{val}(x) = f$ that uses only s-t-paths P with cost $c(P) \leqslant c$. Let $x^{(1)}, x^{(2)}$ denote the restriction of x to the edges in E_1 and E_2, respectively. Furthermore, let P_l denote the longest path that is used by x in G' and let $P_l^{(1)}, P_l^{(2)}$ be the restriction of this path to the edges in E_1 and E_2, respectively. According to the above observation, we may assume that x is positive only on shortest paths in the graph $(G')_y$ for some upgrade profile y, so the cost of the paths used by $x^{(1)}$ in G_1 and by $x^{(2)}$ in G_2 are bounded by $c_1 := c(P_l^{(1)})$ and $c_2 := c(P_l^{(2)})$, respectively, with $c_2 \leqslant c - c_1$. Furthermore, since all flow in G' must first pass G_1 and then G_2, the flows $x^{(1)}$ and $x^{(2)}$ must have flow value f as well. Moreover, the total upgrade costs amount to some values b_1 and $b_2 \leqslant b - b_1$ that arise in G_1 and G_2, respectively. Hence, for the above values of c_1, b_1, and f, we have that $c(x^{(1)}) \geqslant A_{G_1}(c_1, b_1, f)$ and $c(x^{(2)}) \geqslant A_{G_2}(c - c_1, b - b_1, f)$, so

$$A_{G'}(c, b, f) = c(x) = c(x^{(1)}) + c(x^{(2)})$$
$$\geqslant A_{G_1}(c_1, b_1, f) + A_{G_2}(c - c_1, b - b_1, f),$$

as claimed.

Note that, since both $|c_1| \leqslant nC$ and $|c - c_1| \leqslant nC$, we get that

$$-nC \leqslant c_1 \leqslant nC \text{ and } -nC \leqslant c - c_1 \leqslant nC$$
$$\Longleftrightarrow \quad -nC \leqslant c_1 \leqslant nC \text{ and } c - nC \leqslant c_1 \leqslant c + nC$$
$$\Longleftrightarrow \quad \max\{0, c\} - nC \leqslant c_1 \leqslant \min\{0, c\} + nC$$

as in equation (5.2). Hence, it suffices to take the minimum in (5.2) over the given range of c_1.

Parallel composition $G' = G_1 \mid G_2$:

We show that the value $A_{G'}(c, b, f)$ in a graph G' that is the parallel composition of two series-parallel graphs $G_1 = (V_1, E_1)$ and $G_2 = (V_2, E_2)$ is given by

$$A_{G'}(c, b, f) := \min_{\substack{0 \leqslant b_1 \leqslant b \\ 0 \leqslant f_1 \leqslant f}} A_{G_1}(c, b_1, f_1) + A_{G_2}(c, b - b_1, f - f_1). \tag{5.3}$$

"$\leqslant$": Consider some fixed values for b_1 and f_1. The claim clearly holds in the case that $A_{G_1}(c, b_1, f_1) = +\infty$ or $A_{G_2}(c, b - b_1, f - f_1) = +\infty$. Now assume that both terms are finite, i.e., there exist feasible flows $x^{(1)}$ in G_1 and $x^{(2)}$ in G_2 that fulfill the respective bounds on the cost, upgrade cost, and flow value. Hence, by setting $x_e := x_e^{(1)}$ for $e \in E_1$ and $x_e := x_e^{(2)}$ for $e \in E_2$, we get a feasible flow x in G' of flow value $\mathrm{val}(x) = f_1 + (f - f_1) = f$ and upgrade cost at most $b_1 + (b - b_1) = b$ on paths with cost at most c. Thus, we have $A_{G'}(c, b, f) \leqslant c(x) = A_{G_1}(c, b_1, f_1) + A_{G_2}(c, b - b_1, f - f_1)$, which shows the claim.

"$\geqslant$": Let $A_{G'}(c, b, f) < +\infty$. A budget-constrained minimum cost flow x in the graph $G' = G_1 \mid G_2$ uses paths P, each of which is entirely contained either in G_1 or G_2. The budget b is split into two parts b_1 and $b_2 \leqslant b - b_1$ that are incurred in G_1 and G_2, respectively. Similarly, the flow value f of x is composed of some part f_1 in G_1 and some part $f_2 = f - f_1$ in G_2. Since the costs of the used paths are less or equal to c if and only if they are less or equal to c in the respective component in that they are contained, we can define $x^{(1)}$ and $x^{(2)}$ to be the restrictions of x to the edges in E_1, E_2, respectively, and obtain that

$$A_{G'}(c, b, f) = c(x) = c(x^{(1)}) + c(x^{(2)})$$
$$\geqslant A_{G_1}(c, b_1, f_1) + A_{G_2}(c, b - b_1, f - f_1),$$

which shows the claim.

As it was shown in Section 2.3, a decomposition tree of a series-parallel graph G can be computed in $\mathcal{O}(m)$ time and contains $\mathcal{O}(m)$ $(\mathcal{O}(n))$ inner nodes corresponding to parallel compositions (series compositions) and m leaves corresponding to edges in G (cf. (Valdes et al., 1982)). For each node in the decomposition tree corresponding to a series-parallel graph G', we compute the values $A_{G'}(c, b, f)$ for $c \in \{-nC, \dots, nC\}$, $b \in \{0, \dots, B\}$, and $f \in \{0, \dots, mU\}$. By computing these values in a bottom-up manner with respect to the decomposition tree, we can assume that all the values for the subgraphs G_1 and G_2 of a series or parallel composition are already known and are, thus, able to compute each value $A_{G'}(c, b, f)$ in $\mathcal{O}(1)$ time in the case of a single edge, in $\mathcal{O}(nBC)$ time in the case of a series composition, and in $\mathcal{O}(mUB)$ time in the case of a parallel composition. Since an optimal flow cannot be positive on paths with positive cost, the optimal flow value can then be found by selecting the minimum value $A_G(0, B, f)$ for $f \in \{0, \dots, mU\}$. Thus, we get an overall running time of

$$\mathcal{O}(nC \cdot B \cdot mU \cdot (m + n \cdot nBC + m \cdot mUB)) = \mathcal{O}(nmCUB^2 \cdot (n^2C + m^2U))$$

as claimed. By maintaining sets $S_{G'}(c, b, f)$ of pairs (e, x_e) that describe the flow x_e on a specific edge $e \in E$ and merging these sets for those values of c_1, b_1, and f_1 for which the minimum is found in equations (5.2) and (5.3), the optimal flow can be computed alongside with the computation of the values $A_{G'}(c, b, f)$ within the same running time. □

In the following subsection, we concentrate on BCMCFP$_N$ on extension-parallel graphs. This problem turns out to be strongly related to a special variant of the bounded knapsack problem, which will be shown subsequently. Note that, due to the structure of extension-parallel graphs, the number of s-t-paths in such a graph is bounded by the number m of edges. In the pseudo-polynomial-time algorithm introduced in the proof of Theorem 5.5, we are thus able to avoid the minimization over the costs C since we are able to remove paths that have positive costs in a preprocessing step. Hence, we get the following corollary:

Corollary 5.6:
BCMCFP$_N$ is solvable in pseudo-polynomial time $\mathcal{O}(m^3 U^2 B^2)$ on extension-parallel graphs. □

5.4 The Bounded Knapsack Problem with Laminar Cardinality Constraints

While the problem BCMCFP$_N$ was shown to be $\mathcal{NP}$-hard to approximate on series-parallel graphs in Theorem 5.4, we will now consider the case of extension-parallel graphs, for which the problem can be approximated efficiently. As noted above, the problem BCMCFP$_N$ on extension-parallel graphs takes the form of the so called *bounded knapsack problem with laminar cardinality constraints*, which is an extension of the traditional bounded knapsack problem (cf. (Kellerer et al., 2004)) by several *cardinality constraints* of the form $\sum_{i \in I_j} x_i \leqslant \mu_j$ for distinct non-empty subsets $I_1, \ldots, I_h$ of *item types*. The family $\mathcal{J} := \{I_1, \ldots, I_h\}$ is assumed to be *laminar*, i.e., for each pair $(I_1, I_2) \in \mathcal{J} \times \mathcal{J}$, it holds that either $I_1 \cap I_2 = \emptyset$, $I_1 \subset I_2$, or $I_1 \supset I_2$.

Definition 5.7 (Bounded knapsack problem with laminar cardinality constraints (LKP)):

INSTANCE: A set of item types $i \in K := \{1, \ldots, k\}$ together with weights $w_i \in \mathbb{N}_{\geq 0}$, profits $p_i \in \mathbb{N}_{\geq 0}$, and maximum amounts $k_i \in \mathbb{N}_{\geq 0}$ for each $i \in K$ as well as a maximum weight $W \in \mathbb{N}_{\geq 0}$ and a laminar family $\mathfrak{I} := \{I_1, \ldots, I_h\}$ of distinct non-empty subsets of K with maximum amounts $\mu_j \in \mathbb{N}_{\geq 0}$ for $j \in \{1, \ldots, h\}$.

TASK: Find amounts $(x_i)_{i \in K}$ for each item type such that

- $x_i \in \{0, \ldots, k_i\}$ *(bounding constraints)*,

- $\sum_{i \in K} w_i \cdot x_i \leq W$ *(knapsack constraint)*,

- $\sum_{i \in I_j} x_i \leq \mu_j$ for $j \in \{1, \ldots, h\}$ *(cardinality constraints)*, and

- $p(x) := \sum_{i \in K} p_i \cdot x_i$ is maximized. ◁

Note that the problem LKP is identical to the traditional bounded knapsack problem if $h = 0$ and equal to the cardinality-constrained bounded knapsack problem if $h = 1$ and $I_1 = K$ (cf. (Kellerer et al., 2004) for further details on these problems). A similar problem in which the cardinality constraints are replaced by constraints of the form $\sum_{i \in I_j} w_i \cdot x_i \leq \mu_j$ is known as the *arborescent knapsack problem* and was shown to be easy to approximate by Gens and Levner (1980) and Safer and Orlin (1995). However, their algorithms rely on the fact that these new constraints involve partial sums of the sum in the knapsack constraint, which does not apply to our case.

We now show how an instance of $BCMCFP_N$ on an extension-parallel graph can be transformed into an instance of LKP: Let $\mathcal{P} := \{P_1, \ldots, P_k\}$ (with $k \leq m$) denote the set of all s-t-paths in an instance of $BCMCFP_N$. The flow on P_i is bounded by $u(P_i) := \min_{e \in P_i} u_e$ and causes upgrade costs of $b(P_i) := \sum_{e \in P_i} b_e$ and costs of $c(P_i) := \sum_{e \in P_i} c_e$ per unit of flow. The amount of flow on P_i can be seen as the number x_i of items of type i that is packed into the knapsack, where item type i is specified by a maximum amount of $k_i := u(P_i)$ and a weight of $w_i := b(P_i)$ as well as a profit of $p_i := -c(P_i)$ per item. The knapsack constraint is then equivalent to constraint (5.1c) when defining $W := B$. Furthermore, the flows on all paths that have some edge $e \in E$ in common must fulfill the capacity constraint of this edge. This requirement is modeled by introducing a cardinality constraint $\sum_{i \in I_j} x_j \leq \mu_j$ for each edge e that is contained in multiple paths by setting $\mu_j := u_e$ and defining I_j to be the set of indices of all item types i that correspond to paths P_i containing e. Note that these sets fulfill the laminarity property because of the special structure of extension-parallel graphs.

Note that each s-t-path P in the underlying graph of a given instance of $BCMCFP_N$ is determined by at least one *characteristic edge* $e = (v, w)$ that is only contained in P and is, thus, reachable by exactly one path $P^{(1)}$ from the source s and connected to the sink t by exactly one path $P^{(2)}$. Thus, for each edge $e \in E$, we can check if it corresponds to a characteristic edge in $\mathcal{O}(n)$ time by checking if $\delta^-(v) \leqslant 1$ for each $v \in P^{(1)}$ and if $\delta^+(v) \leqslant 1$ for each $v \in P^{(2)}$. Thus, we can determine all s-t-paths together with their flow, capacity, costs, and upgrade costs in $\mathcal{O}(nm)$ time. Hence, the above transformation and the corresponding back transformation can both be performed in $\mathcal{O}(nm)$ time. An exemplary transformation of an instance of $BCMCFP_N$ to an instance of LKP is shown in Figure 5.2.

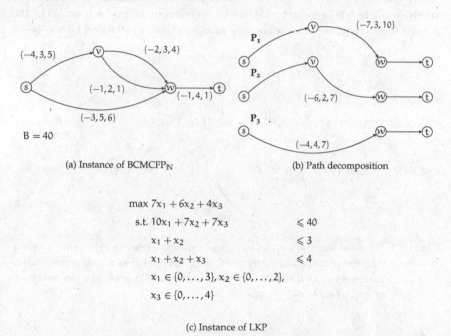

(a) Instance of $BCMCFP_N$ (b) Path decomposition

$$\max 7x_1 + 6x_2 + 4x_3$$
$$\text{s.t. } 10x_1 + 7x_2 + 7x_3 \leqslant 40$$
$$x_1 + x_2 \leqslant 3$$
$$x_1 + x_2 + x_3 \leqslant 4$$
$$x_1 \in \{0, \dots, 3\}, x_2 \in \{0, \dots, 2\},$$
$$x_3 \in \{0, \dots, 4\}$$

(c) Instance of LKP

Figure 5.2: An instance of $BCMCFP_N$ on an extension-parallel graph (upper left), the path decomposition (upper right), and the corresponding instance of LKP (bottom) as an integer linear program. The label on each edge and path denotes its cost, capacity, and upgrade cost, respectively.

In the rest of this section, we focus on LKP, but we always state the corresponding results for $BCMCFP_N$ on extension-parallel graphs that can be obtained by using the above transformation explicitly.

Without loss of generality, for each $i \in K$, we assume that the values p_i are positive (since item types with a non-positive profit can be neglected) and that the solution that contains k_i items of type i and no items of other types is feasible (since we can otherwise reduce the value k_i appropriately).

Finally, note that the laminar family $\mathcal{I} = \{I_1, \ldots, I_h\}$ can be represented by a forest, in which the nodes v_i correspond to the sets $I_i \in \mathcal{I}$ and in which a node v_i is a successor of node v_j if and only if $I_i \subset I_j$. This implies that v_i is a direct child of v_j if and only if $I_i \subset I_j$ and there is no $I_k \in \mathcal{I}$ such that $I_i \subset I_k \subset I_j$. Moreover, by adding artificial sets I' with large maximum amounts $\mu' := \sum_{i \in K} k_i$, we can turn this forest into a binary tree (cf. Figure 5.3). This, however, implies that the number h of cardinality constraints (which is bounded by the number of nodes in the tree) is bounded by the number k of item types (which bounds the number of leaves in the tree) by $h \leqslant 2k - 1$.

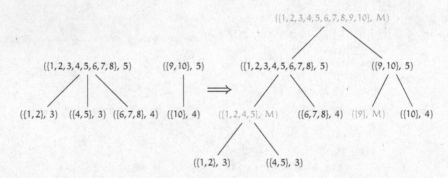

Figure 5.3: A laminar family of sets and its representation as a forest (left) as well as a equivalent representation as a binary tree (right). The label of each node corresponds to the respective set I_j and the maximum amount μ_j. The artificial nodes are shown in gray, where $M := \sum_{i \in K} k_i$.

5.4.1 A Polynomial-Time Approximation Scheme

In the following, let $p(x) := \sum_{i \in K} p_i \cdot x_i$ and $w(x) := \sum_{i \in K} w_i \cdot x_i$ denote the profit generated and the weight used by a solution x of LKP, respectively. Although, to the best of our knowledge, the problem LKP has not been considered in literature so far, we are able to adopt several results that are valid for the traditional (bounded) knapsack problem to the case of LKP. In this subsection, we derive a 2-approximation algorithm for the problem that will subsequently incorporated into a polynomial-time approximation scheme. To this end, we need three auxiliary lemmas:

Lemma 5.8:
An optimal solution of the LP-relaxation of LKP can be computed in $\mathcal{O}(k^2 \log^2 k)$ time.

Proof: Clearly, the LP-relaxation of LKP can be transformed into an equivalent instance of $\text{BCMCFP}_\mathbb{R}$ and vice versa using the same procedure as above. According to Corollary 4.23 in Section 4.4, we can solve $\text{BCMCFP}_\mathbb{R}$ in $\mathcal{O}(m^2 \log^2 m)$ time on series-parallel graphs (and, thus, on extension-parallel graphs as well). By transforming the obtained solution of $\text{BCMCFP}_\mathbb{R}$ to a solution of the LP-relaxation of LKP, the claim follows. □

Lemma 5.9:
Each optimal solution of the LP-relaxation of LKP can be turned into an optimal solution with at most two fractional values in $\mathcal{O}(k^2)$ time.

Proof: According to Theorem 4.7 on page 47, we can transform each optimal solution of $\text{BCMCFP}_\mathbb{R}$ into an optimal basic feasible solution in $\mathcal{O}(m^2)$ time. According to Corollary 4.6 on page 46, the flow on each edge is integral except for the edges on at most one cycle $C(\bar{e})$. Let $v_1 \in C(\bar{e})$ and $v_2 \in C(\bar{e})$ denote the nodes with the lowest and the highest index in a topological sorting of the edge set V, respectively. The cycle $C(\bar{e})$ then consists of two parallel path segments between v_1 and v_2, which contain the characteristic edges of two s-t-paths, that, consequently, are the only s-t-paths with fractional flow. □

Lemma 5.10:
If an optimal solution x^{LP} of the LP-relaxation of LKP contains exactly two fractional values x_j^{LP} and x_l^{LP} with $w_j \leqslant w_l$, then the vector $(x_1, \ldots, x_k)$ with $x_j := \left\lceil x_j^{\text{LP}} \right\rceil$, $x_l := \left\lfloor x_l^{\text{LP}} \right\rfloor$, and $x_i := x_i^{\text{LP}}$ for $i \in K \setminus \{j, l\}$ is a feasible solution of LKP.

Proof: Let $q_j, q_l \in \mathbb{N}_{\geqslant 0}$ and $r_j, r_l \in (0, 1)$ such that $x_j^{\text{LP}} = q_j + r_j$ and $x_l^{\text{LP}} = q_l + r_l$. Note that we can assume without loss of generality that $r_j + r_l = 1$: If $r_j + r_l \neq 1$, there cannot be a cardinality constraint that contains any of the two variables and that is fulfilled with equality since x_j^{LP} and x_l^{LP} are the only two fractional values and the right-hand side of each cardinality constraint is integral. Note that both $w_j > 0$ and $w_l > 0$ without loss of generality since we could otherwise improve the solution x^{LP} by increasing the corresponding variable with zero weight. Therefore, as in the case of the traditional bounded knapsack problem, we can replace ε elements of item type j with $\frac{w_j}{w_l} \cdot \varepsilon$ elements of item type l if $\frac{p_j}{w_j} \leqslant \frac{p_l}{w_l}$ or vice versa if $\frac{p_j}{w_j} \geqslant \frac{p_l}{w_l}$ without losing profit or violating the knapsack constraint. For a suitable choice of ε, the resulting solution will either have at most one fractional value left or will fulfill a cardinality constraint with equality that was fulfilled with strict inequality before. In the latter case, the claim follows since the right-hand side of each cardinality constraint is integral.

Hence, as $x_j = \left\lceil x_j^{LP} \right\rceil = x_j^{LP} + (1 - r_j)$ and $x_l = \left\lfloor x_l^{LP} \right\rfloor = x_l^{LP} - r_l$, we get that the new solution x does not violate the cardinality constraints. It remains to show that the knapsack condition is still fulfilled. But since $w_j \leqslant w_l$, we get that

$$
\begin{aligned}
w_j \cdot x_j + w_l \cdot x_l &= w_j \cdot \left\lceil x_j^{LP} \right\rceil + w_l \cdot \left\lfloor x_l^{LP} \right\rfloor \\
&= w_j \cdot (x_j^{LP} + (1 - r_j)) + w_l \cdot (x_l^{LP} - r_l) \\
&= w_j \cdot x_j^{LP} + w_l \cdot x_l^{LP} + w_j \cdot (1 - r_j) - w_l \cdot r_l \\
&\leqslant w_j \cdot x_j^{LP} + w_l \cdot x_l^{LP} + w_l \cdot (1 - r_j - r_l) \\
&= w_j \cdot x_j^{LP} + w_l \cdot x_l^{LP}.
\end{aligned}
$$

Since the other values remain unchanged and x^{LP} fulfills the knapsack condition as well, the feasibility of x follows. $\qquad\square$

Using the results of the above lemmas, we obtain the following 2-approximation algorithm for LKP.

Theorem 5.11:
There is a 2-approximation algorithm for LKP with a running time in $\mathcal{O}(k^2 \cdot \log^2 k)$.

Proof: Consider the following algorithm:
1: Compute an optimal solution x^{LP} of the LP-relaxation in $\mathcal{O}(k^2 \cdot \log^2 k)$ time (Lemma 5.8).
2: Turn x^{LP} into a solution with at most two fractional values in $\mathcal{O}(k^2)$ time (Lemma 5.9).
3: Let F be the the set of fractional variables in x^{LP}.

4: **if** $F = \emptyset$ **then**
5: $\quad x := x^{LP}$.

6: **else if** $F = \{x_l\}$ **then**
7: $\quad x_l := \left\lfloor x_l^{LP} \right\rfloor$ and $x_i := x_i^{LP}$ for $i \in K \setminus \{l\}$.
8: $\quad$ **if** $p(x) < p_l$ **then**
9: $\quad\quad x_l := 1$ and $x_i := 0$ for $i \in K \setminus \{l\}$.

10: **else if** $F = \{x_j, x_l\}$ with $w_j \leqslant w_l$ **then**

11:     $x_j := \left\lceil x_j^{LP} \right\rceil$, $x_l := \left\lfloor x_l^{LP} \right\rfloor$, and $x_i := x^{LP}$ for $i \in K \setminus \{j, l\}$.

12:     **if** $p(x) < p_l$ **then**

13:        $x_l := 1$ and $x_i := 0$ for $i \in K \setminus \{l\}$.

    return x.

The feasibility of the constructed solution follows from Lemma 5.10. Let x^* denote the optimal solution of the given instance of LKP. If $F = \emptyset$, the solution x computed by the algorithm is obviously optimal. Otherwise, if $F = \{x_l\}$, we have

$$
\begin{aligned}
p(x^*) = \sum_{i \in K} p_i \cdot x_i^* &\leqslant \sum_{i \in K} p_i \cdot x_i^{LP} \leqslant \sum_{i \in K \setminus \{l\}} p_i \cdot x_i^{LP} + p_l \cdot \left(\left\lfloor x_l^{LP} \right\rfloor + 1 \right) \\
&= \left(\sum_{i \in K \setminus \{l\}} p_i \cdot x_i^{LP} + p_l \cdot \left\lfloor x_l^{LP} \right\rfloor \right) + p_l \\
&\leqslant 2 \cdot \sum_{i \in K} p_i \cdot x_i = 2p(x).
\end{aligned} \tag{5.4}
$$

Finally, if $F = \{x_j, x_l\}$, we get that

$$
\begin{aligned}
p(x^*) = \sum_{i \in K} p_i \cdot x_i^* &\leqslant \sum_{i \in K} p_i \cdot x_i^{LP} \\
&\leqslant \sum_{i \in K \setminus \{j, l\}} p_i \cdot x_i^{LP} + p_j \cdot \left\lceil x_j^{LP} \right\rceil + p_l \cdot \left(\left\lfloor x_l^{LP} \right\rfloor + 1 \right) \\
&= \left(\sum_{i \in K \setminus \{j, l\}} p_i \cdot x_i^{LP} + p_j \cdot \left\lceil x_j^{LP} \right\rceil + p_l \cdot \left\lfloor x_l^{LP} \right\rfloor \right) + p_l \\
&\leqslant 2 \cdot \sum_{i \in K} p_i \cdot x_i = 2p(x).
\end{aligned} \tag{5.5}
$$

Hence, the claim follows.   □

As motivated above, the results of Theorem 5.11 can be immediately applied to the problem BCMCFP$_N$. Recall that $c(x^*) \leqslant 0$ for any optimal solution x^* to BCMCFP$_N$, so we need to handle BCMCFP$_N$ as a maximization problem rather than a minimization problem when speaking about approximation algorithm.

Corollary 5.12:
There is a 2-approximation algorithm for BCMCFP$_N$ on extension-parallel graphs with a running time in $\mathcal{O}(m^2 \cdot \log^2 m)$.   □

Similarly as in the case of the traditional bounded knapsack problem (cf. (Kellerer et al., 2004)), we are able to use the above 2-approximation algorithm in order to obtain a PTAS for LKP:

Theorem 5.13:

There is a PTAS for LKP that runs in $\mathcal{O}\left(\left(\frac{k}{\varepsilon}\right)^{\lceil\frac{1}{\varepsilon}\rceil-2} \cdot \left(\frac{k}{\varepsilon} + k^2\log^2 k\right) \right)$ time.

Proof: As noted above, the proof of the theorem works similarly as in the case of the traditional bounded knapsack problem. For a suitable choice of a parameter q, the idea of the PTAS is to guess the q elements with the largest profit contained in the knapsack and to fill the remaining capacity with items of smaller profit using the 2-approximation algorithm. In order to do so, we extend the instance of LKP to contain $\min\{k_i, q\}$ copies of each item type and let $K' := \{(i, l) : i \in K, l \in \{1, \ldots, \min\{k_i, q\}\}\}$ denote the extended set of item types.

1: $q := \min\left\{\lceil\frac{1}{\varepsilon}\rceil - 2, \sum_{i\in K} k_i\right\}$.

2: $x_i := 0$ for $i \in K$.

3: **for all** $L \subset K'$ with $0 < |L| < q$ **do**

4:     **if** $\sum_{(i,l)\in L} w_i \leq W$ and $|\{(i, l) \in L : i \in I_j\}| \leq \mu_j$ for each $j \in \{1, \ldots, h\}$ **then**

5:         **if** $\sum_{(i,l)\in L} p_i > p(x)$ **then**

6:             $x_i := |\{(i, l) \in L\}|$ for $i \in K$.

7: **for all** $L \subseteq K'$ with $|L| = q$ **do**

8:     **if** $\sum_{(i,l)\in L} w_i \leq W$ and $|\{(i, l) \in L : i \in I_j\}| \leq \mu_j$ for each $j \in \{1, \ldots, h\}$ **then**

9:         $S := \{i \in K : p_i \leq p_{i'}$ for all $(i', l) \in L\}$.

10:         $W' := W - \sum_{(i,l)\in L} w_i$.

11:         $\mu_j' := \mu_j - |\{(i, l) \in L : i \in I_j\}|$ for $j \in \{1, \ldots, h\}$.

12:         $k_i' := k_i - |\{(i, l) \in L\}|$ for $i \in S$.

13:         Let x' denote the result of the 2-approximation algorithm for LKP applied to the instance with item types S, maximum weight W', maximum amounts k_i' for each item type $i \in S$ and maximum amounts μ_j' for each cardinality constraint $j \in \{1, \ldots, h\}$.

14:         **if** $\sum_{(i,l)\in L} p_i + \sum_{i\in S} p_i \cdot x_i' > p(x)$ **then**

15:             $x_i' := 0$ for $i \in K \setminus S$.

16:             $x_i := x_i' + |\{(i, l) \in L\}|$ for each $i \in K$.

    return x.

The proof of correctness works analogously to the case of the traditional bounded knapsack problem as shown in (Kellerer et al., 2004): First consider the case that the optimal solution consists of less than q items. Clearly, the optimal solution will then be found by the first loop. Otherwise, let $L^* \subseteq K'$ denote the set of those q items that contribute the highest profit to the optimal solution x^*. Note that L^* will be considered during the execution of the second loop. If $q = \sum_{i\in K} k_i$, we clearly obtain an optimal solution in this case. Else, note that the optimal solution consists of the items in L^* and an optimal solution x_S^* to the subproblem considered during the above algorithm with the item types $S := \{i \in K : p_i \leqslant p_{i'} \text{ for all } (i', l) \in L^*\}$. The algorithm then computes a 2-approximation x_S' on S which consequently fulfills $p(x_S') \geqslant \frac{1}{2} \cdot p(x_S^*)$. If x is the solution returned by the algorithm, we distinguish between the following two cases according to (Kellerer et al., 2004):

Case 1: $\sum_{(i,l) \in L^*} p_i \geqslant \frac{q}{q+2} \cdot p(x^*)$. In this case, we get that

$$
\begin{aligned}
p(x) = \sum_{(i,l) \in L^*} p_i + p(x_S') &\geqslant \sum_{(i,l) \in L^*} p_i + \frac{1}{2} \cdot p(x_S^*) \\
&= \sum_{(i,l) \in L^*} p_i + \frac{1}{2} \cdot \left(p(x^*) - \sum_{(i,l) \in L^*} p_i \right) \\
&= \frac{1}{2} \cdot \left(p(x^*) + \sum_{(i,l) \in L^*} p_i \right) \geqslant \frac{1}{2} \cdot \left(p(x^*) + \frac{q}{q+2} \cdot p(x^*) \right) \\
&= \frac{q+1}{q+2} \cdot p(x^*).
\end{aligned}
$$

Case 2: $\sum_{(i,l) \in L^*} p_i < \frac{q}{q+2} \cdot p(x^*)$. In this case, there is at least one item in L^* with profit less than $\frac{1}{q+2} \cdot p(x^*)$, which implies that x_S^* uses only items with profit less than $\frac{1}{q+2} \cdot p(x^*)$ as well. In particular, it holds that $p_l < \frac{1}{q+2} \cdot p(x^*)$ for the item type l in equations (5.4) and (5.5) in the proof of Theorem 5.11, such that

$$
p(x_S^*) \leqslant p(x_S') + p_l \leqslant p(x_S') + \frac{1}{q+2} \cdot p(x^*),
$$

which implies that

$$
\begin{aligned}
p(x^*) = \sum_{(i,l) \in L^*} p_i + p(x_S^*) &\leqslant \sum_{(i,l) \in L^*} p_i + p(x_S') + \frac{1}{q+2} \cdot p(x^*) \\
&\leqslant p(x) + \frac{1}{q+2} \cdot p(x^*).
\end{aligned}
$$

Hence, we get that

$$
p(x) \geqslant p(x^*) - \frac{1}{q+2} \cdot p(x^*) = \frac{q+1}{q+2} \cdot p(x^*).
$$

Thus, in any of the two cases, it holds that $p(x) \geqslant \frac{q+1}{q+2} \cdot p(x^*)$. Moreover, since $\frac{q+1}{q+2}$ is increasing with q, we further get that

$$
p(x) \geqslant \frac{q+1}{q+2} \cdot p(x^*) = \frac{\lceil \frac{1}{\varepsilon} \rceil - 1}{\lceil \frac{1}{\varepsilon} \rceil} \cdot p(x^*) \geqslant \frac{\frac{1}{\varepsilon} - 1}{\frac{1}{\varepsilon}} \cdot p(x^*) = (1 - \varepsilon) \cdot p(x^*),
$$

which shows the correctness of the algorithm.

The running time of the algorithm is dominated by the determination of the subsets L of K', whose number is given by $\sum_{i=1}^{q} \binom{k \cdot q}{i} \in \mathcal{O}((k \cdot q)^q)$. For each such subset with cardinality q, we need to check if it allows a feasible solution and, if so, construct the corresponding subproblem in $\mathcal{O}(h \cdot q + k) = \mathcal{O}(k \cdot q)$ time (recall that $h \in \mathcal{O}(k)$). The algorithm then computes a 2-approximation for this subproblem in $\mathcal{O}(k^2 \log^2 k)$ time, which results in a total running time of

$$\mathcal{O}\left((k \cdot q)^q \cdot \left(k \cdot q + k^2 \log^2 k\right)\right) = \mathcal{O}\left(\left(\frac{k}{\varepsilon}\right)^{\lceil \frac{1}{\varepsilon} \rceil - 2} \cdot \left(\frac{k}{\varepsilon} + k^2 \log^2 k\right)\right),$$

which completes the proof. □

Corollary 5.14:
There is a PTAS for BCMCFP$_N$ on extension-parallel graphs with a running time in $\mathcal{O}\left(\left(\frac{m}{\varepsilon}\right)^{\lceil \frac{1}{\varepsilon} \rceil - 2} \cdot \left(\frac{m}{\varepsilon} + m^2 \log^2 m\right)\right).$ □

5.4.2 Fully Polynomial-Time Approximation Schemes

Note that the algorithm in Theorem 5.13 yields no FPTAS since its running time is not polynomial in $\frac{1}{\varepsilon}$. In this subsection, we develop two FPTASs for the problem under several restrictions that are based on different approaches. We first show that there is an FPTAS if the maximum amounts k_i are polynomially bounded. We then show that there is a *bicriteria* FPTAS for LKP and BCMCFP$_N$ on extension-parallel graphs that computes a solution x which fulfills $p(x) \geqslant (1 - \varepsilon) \cdot p(x^*)$ and $w(x) \leqslant (1 + \varepsilon) \cdot w(x^*)$ with respect to the optimal solution x^*. As a by-product, we also show that an ε-approximate pareto frontier of LKP and BCMCFP$_N$ on extension-parallel graphs can be computed efficiently.

Theorem 5.15:
There is an FPTAS for LKP that runs in $\mathcal{O}\left(k \cdot \bar{\mu}^4 \cdot \frac{1}{\varepsilon^2}\right)$ time if the maximum number $\bar{\mu}$ of items in an optimal solution is polynomially bounded.

Proof: Similar to Theorem 5.5, we develop an exact pseudo-polynomial-time algorithm. Afterwards, we show that we can scale the profits p_i for $i \in K$ such that the algorithm becomes an FPTAS.

In the following, let $\bar{\mu} := \sum_{i \in K} k_i$ denote an upper bound on the maximum number of items in the knapsack. Similarly, let $P := \bar{\mu} \cdot \max_{i \in K} p_i$ denote an upper bound on the maximum profit of a feasible solution. As we have shown on page 92, we can represent the laminar family $\mathcal{J}$ as a binary tree (cf. Figure 5.3). Furthermore, we may assume that there are no sets in $\mathcal{J}$ with cardinality one since a cardinality constraint of the form $x_i \leqslant \mu_j$ can be modeled by updating k_i to $\min\{k_i, \mu_j\}$. Thus, considering the item types as leaves and the sets as inner nodes, we obtain a binary tree T with k leaves and $k - 1$ inner nodes.

Let $A_{T'}(\mu, p)$ denote the minimum weight that is needed in order to create a total profit of at least p with the item types that are contained in the subtree T' of T while using at most μ such items. For the case that T' is a leaf of T, i.e., T' corresponds to a single item type $i \in K$, we have

$$A_{T'}(\mu, p) := \begin{cases} w_i \cdot \left\lceil \frac{p}{p_i} \right\rceil, & \text{if } \left\lceil \frac{p}{p_i} \right\rceil \leqslant \min\{\mu, k_i\}, \\ \infty, & \text{else,} \end{cases}$$

since we need at least $\left\lceil \frac{p}{p_i} \right\rceil$ items of type i in order to create a profit of p or more, which is only possible if the bounds μ and k_i allow it. Similarly, for some subtree T' of T whose root is some inner node of T that corresponds to a set I_j and whose two children are the roots of the subtrees T_1 and T_2, we get that

$$A_{T'}(\mu, p) := \begin{cases} \min_{\substack{0 \leqslant \mu_1 \leqslant \mu \\ 0 \leqslant p_1 \leqslant p}} A_{T_1}(\mu_1, p_1) + A_{T_2}(\mu - \mu_1, p - p_1), & \text{if } \mu \leqslant \mu_j \\ \infty, & \text{else,} \end{cases}$$

following similar arguments as used in the proof of Theorem 5.5. Since T contains $2k - 1$ nodes and there are $\overline{\mu}$ and P possible values for μ and p, respectively, each value $A_{T'}(\mu, p)$ can be computed in $\mathcal{O}(\overline{\mu} \cdot P)$ time, so we get a total running time of $\mathcal{O}(k \cdot (\overline{\mu} \cdot P) \cdot (\overline{\mu} \cdot P)) = \mathcal{O}\left(k \cdot \overline{\mu}^2 \cdot P^2\right)$ for computing all values $A_{T'}(\mu, p)$ with $0 \leqslant \mu \leqslant \overline{\mu}$ and $0 \leqslant p \leqslant P$.

We now show how we can use this procedure in order to obtain an FPTAS for LKP. Therefore, let $\varepsilon \in (0, 1)$ be a value determining the quality of the approximation. Furthermore, let x^* denote an optimal solution and let x^A be the solution obtained by the 2-approximation algorithm introduced in Theorem 5.11 such that $p(x^A) \geqslant \frac{1}{2} \cdot p(x^*)$.

Suppose that x is the solution obtained by the above procedure using the profits $\widetilde{p}_i := \left\lfloor \frac{p_i}{M} \right\rfloor$ for $i \in K$ with $M := \frac{\varepsilon \cdot p(x^A)}{\overline{\mu}}$. Using the results and definitions so far, we get that

$$p(x) = \sum_{i \in K} p_i \cdot x_i \geqslant \sum_{i \in K} M \cdot \left\lfloor \frac{p_i}{M} \right\rfloor \cdot x_i \geqslant \sum_{i \in K} M \cdot \left\lfloor \frac{p_i}{M} \right\rfloor \cdot x_i^*$$

$$\geqslant \sum_{i \in K} M \cdot \left(\frac{p_i}{M} - 1\right) \cdot x_i^* = \sum_{i \in K} (p_i - M) \cdot x_i^* = p(x^*) - M \cdot \sum_{i \in K} x_i^*$$

$$\geqslant p(x^*) - M \cdot \overline{\mu} = p(x^*) - \varepsilon \cdot p(x^A)$$

$$\geqslant p(x^*) - \varepsilon \cdot p(x^*) = (1 - \varepsilon) \cdot p(x^*).$$

Hence, the procedure computes a solution with profit at least $(1 - \varepsilon) \cdot p(x^*)$. Furthermore, each solution $\widetilde{x}$ of an instance of LKP with the profits $\widetilde{p}_i$ satisfies

$$\widetilde{p}(\widetilde{x}) := \sum_{i \in K} \widetilde{p}_i \cdot \widetilde{x}_i = \sum_{i \in K} \left\lfloor \frac{p_i}{M} \right\rfloor \cdot \widetilde{x}_i \leqslant \frac{1}{M} \cdot \sum_{i \in K} p_i \cdot \widetilde{x}_i \leqslant \frac{1}{M} \cdot p(x^*)$$

$$\leqslant \frac{1}{M} \cdot 2p(x^A) = \frac{2\overline{\mu}}{\varepsilon},$$

so the maximum possible total profit $\widetilde{P}$ using the item profits $\widetilde{p}_i$ is polynomially bounded by $\frac{2\overline{\mu}}{\varepsilon}$. Using this bound in the algorithm, we, thus, obtain a total running time of $\mathcal{O}\left(k \cdot \overline{\mu}^4 \cdot \frac{1}{\varepsilon^2}\right)$, which is polynomial in the encoding size of the instance and $\frac{1}{\varepsilon}$, so the algorithm is in fact an FPTAS. $\square$

Corollary 5.16:
There is an FPTAS for $\text{BCMCFP}_\mathbb{N}$ on extension-parallel graphs with a running time in $\mathcal{O}\left(m \cdot F^4 \cdot \frac{1}{\varepsilon^2}\right)$ if the maximum flow value F is polynomially bounded. $\square$

We now show that we can compute an ε-approximate pareto frontier for LKP efficiently by using an approach based on a framework which was introduced by Papadimitriou and Yannakakis (2000) and later extended to the case of minimization objectives by Mittal and Schulz (2013). This in turn yields a bricriteria FPTAS for LKP and $\text{BCMCFP}_\mathbb{N}$ on extension-parallel graphs, as it will be shown in the following. Although the running time of the following algorithms are far away from practical importance, the results show that there is a bicriteria fully polynomial-time approximation scheme when considering $\text{BCMCFP}_\mathbb{N}$ on extension-parallel graphs. This is contrary to the case of series-parallel graphs in general, for which $\text{BCMCFP}_\mathbb{N}$ is $\mathcal{NP}$-hard to approximate according to Theorem 5.4.

In the following, for some instance of LKP, let X denote the set of all integral solutions that satisfy both the cardinality and the bounding constraints (but not necessarily the knapsack constraint). We repeat the definition of an ε-approximate pareto frontier, tailored to the case of LKP:

Definition 5.17 (ε-Approximate Pareto Frontier for LKP):
For an instance of LKP and some value $\varepsilon \in (0, 1)$, a subset $\mathcal{P}(\varepsilon) \subseteq X$ is called an ε-approximate pareto frontier (for LKP) if, for all $x \in X$, there is a point $x_P \in \mathcal{P}(\varepsilon)$ with

$$(1 + \varepsilon) \cdot p(x_P) \geqslant p(x) \qquad \text{and} \qquad w(x_P) \leqslant (1 + \varepsilon) \cdot w(x). \qquad \triangleleft$$

Intuitively, an ε-approximate pareto frontier contains a "sufficiently good" approximation of the real pareto frontier with respect to the precision parameter ε. Since we are ultimately interested in a bicriteria FPTAS for LKP, we assume that $\varepsilon \in (0,1)$, which also simplifies some of the upcoming results. As shown by Papadimitriou and Yannakakis (2000), for each multi-objective optimization problem, there always exists an ε-approximate pareto frontier of polynomial size if each of the objective functions can be evaluated in polynomial time (which is clearly true in our linear setting). Moreover, for a large class of such problems, such an ε-approximate pareto frontier can also be determined in polynomial time. As it will be shown in the following, the problem LKP is contained in this class, which both yields an efficient algorithm for the creation of $\mathcal{P}(\varepsilon)$ and a bicriteria FPTAS for LKP.

In the following, let $\varepsilon' := 1 - \frac{1}{\sqrt{1+\varepsilon}}$ and $\varepsilon'' := \sqrt{1+\varepsilon} - 1$ be two adjusted precision parameters with $(1-\varepsilon')^2 = \frac{1}{1+\varepsilon}$ and $(1+\varepsilon'')^2 = (1+\varepsilon)$. Note that $\frac{1}{\varepsilon'} \leqslant \frac{2+\sqrt{2}}{\varepsilon}$ and $\frac{1}{\varepsilon''} \leqslant \frac{1+\sqrt{2}}{\varepsilon}$ for $\varepsilon \in (0,1)$, so we get that both $\frac{1}{\varepsilon'} \in \mathcal{O}\left(\frac{1}{\varepsilon}\right)$ and $\frac{1}{\varepsilon''} \in \mathcal{O}\left(\frac{1}{\varepsilon}\right)$. One essential ingredient for the creation of the ε-approximate pareto frontier is the so called *gap problem*, which can be defined as follows:

Definition 5.18 (Gap-Problem for LKP):

INSTANCE: A pair (ρ, ω) and a value $\varepsilon \in (0,1)$.

QUESTION: Either return a solution $x \in X$ with $p(x) \geqslant \rho$ and $w(x) \leqslant \omega$ or state that there is no $x \in X$ with $p(x) \geqslant (1+\varepsilon'') \cdot \rho$ and $w(x) \leqslant (1-\varepsilon') \cdot \omega$ for $\varepsilon'' = \sqrt{1+\varepsilon} - 1$ and $\varepsilon' = 1 - \frac{1}{\sqrt{1+\varepsilon}}$. ◁

In particular, note that the gap-problem may behave nondeterministically in case that both there is a solution $x \in X$ with $p(x) \geqslant \rho$ and $w(x) \leqslant \omega$, and there is no solution x' with $p(x') \geqslant (1+\varepsilon'') \cdot \rho$ and $w(x') \leqslant (1-\varepsilon') \cdot \omega$. The following lemma shows that the gap-problem can be solved in polynomial time in the case of LKP:

Lemma 5.19:

For a given pair (ρ, ω) and a value $\varepsilon \in (0,1)$, the gap-problem for LKP can be solved in $\mathcal{O}\left(\frac{k^5}{\varepsilon^4} \cdot \log^5 \overline{k}\right)$ time, where $\overline{k} := \max_{i \in K} k_i$.

Proof: The proof of the lemma is established in three steps: First, we show that we can transform each instance I of LKP into an equivalent instance I' that uses binary variables only and that this transformation increases the number of variables only by a factor of $\mathcal{O}(\log \overline{k})$. Second, we will use this transformed instance I' in order to derive conditions that reduce the gap-problem to an instance I'' in which the parameters are scaled to polynomial size. Finally, we show that a solution to this scaled instance I'' can be found in polynomial time.

Consider an instance I of LKP and let $\bar{l}_i := \lfloor \log_2(k_i) \rfloor$ for each $i \in K$. We replace each variable x_i by $\bar{l}_i + 1$ variables $x_{i,l} \in \{0, 1\}$ for $l \in \{0, \ldots, \bar{l}_i\}$ with the following meaning: For each $l \in \{0, \ldots, \bar{l}_i - 1\}$, variable $x_{i,l} = 1$ if and only if $\eta_{i,l} := 2^l$ items of type i are packed into the knapsack while variable $x_{i,\bar{l}_i} = 1$ if and only if the remaining $\eta_{i,\bar{l}_i} := k_i - \sum_{l \in \{0,\ldots,\bar{l}_i-1\}} 2^l = k_i - 2^{\bar{l}_i} + 1$ items of type i are used. Consequently, the profit $p_{i,l}$ and weight $w_{i,l}$ of item $x_{i,l}$ is $\eta_{i,l}$ times the profit and weight of item type i, respectively. Analogously, each cardinality constraint of the form $\sum_{i \in I_j} x_i \leqslant \mu_j$ is replaced by a new constraint of the form $\sum_{i \in I_j} \sum_{l \in \{0,\ldots,\bar{l}_i\}} \eta_{i,l} \cdot x_{i,l} \leqslant \mu_j$. Clearly, each amount x_i of the original instance of LKP can be represented by suitable choices of the values $x_{i,l}$, $l \in \{0, \ldots, \bar{l}_i\}$, and vice versa. Note that the number k' of items $K' = \{(i, l) : i \in K, l \in \{0, \ldots, \bar{l}_i\}\}$ increases by a logarithmic factor of $\mathcal{O}(\log \bar{k})$. In the following, we can, thus, assume that the underlying problem uses binary variables only without loss of generality.

Suppose that the gap-problem is called for some pair (ρ, ω). Let $M_p := \lceil \frac{k'}{\varepsilon''} \rceil$ and $M_w := \lceil \frac{k'}{\varepsilon'} \rceil$, where ε'' and ε' are given as in Definition 5.18. Furthermore, let $\bar{p}_{i,l} := \min \left\{ \lfloor \frac{p_{i,l} \cdot M_p}{\rho} \rfloor, M_p \right\}$ and $\bar{w}_{i,l} := \lceil \frac{w_{i,l} \cdot M_w}{\omega} \rceil$ for $(i, l) \in K'$ denote scaled profits and weights. Accordingly, we define the scaled profit- and weight-functions as $\bar{p}(x) = \sum_{(i,l) \in K'} \bar{p}_{i,l} \cdot x_{i,l}$ and $\bar{w}(x) = \sum_{(i,l) \in K'} \bar{w}_{i,l} \cdot x_{i,l}$, respectively. In the following two claims, we show that we can solve the gap-problem by investigating the value of the scaled functions:

Claim: If $\bar{p}(x) \geqslant M_p$, it holds that $p(x) \geqslant \rho$. Moreover, if $p(x) \geqslant \rho \cdot (1 + \varepsilon'')$, we get that $\bar{p}(x) \geqslant M_p$.

Proof: Suppose that $\bar{p}(x) \geqslant M_p$. For the profit $p(x)$, we get that

$$p(x) = \sum_{(i,l) \in K'} p_{i,l} \cdot x_{i,l} = \frac{\rho}{M_p} \cdot \sum_{(i,l) \in K'} \frac{p_{i,l} \cdot M_p}{\rho} \cdot x_{i,l}$$

$$\geqslant \frac{\rho}{M_p} \cdot \sum_{(i,l) \in K'} \bar{p}_{i,l} \cdot x_{i,l} = \frac{\rho}{M_p} \cdot \bar{p}(x) \geqslant \frac{\rho}{M_p} \cdot M_p = \rho.$$

Now suppose that $p(x) \geqslant \rho \cdot (1 + \varepsilon'')$. Let $J \subseteq K'$ be the set of all $(i, l) \in K'$ with $x_{i,l} = 1$. If there is some $(i, l) \in J$ with $\bar{p}_{i,l} = M_p$, the claim obviously follows. Otherwise, if $\bar{p}_{i,l} < M_p$ for all $(i, l) \in J$, we get the following lower bound for $\bar{p}(x)$:

$$\bar{p}(x) = \sum_{(i,l) \in K'} \bar{p}_{i,l} \cdot x_{i,l} = \sum_{(i,l) \in J} \left\lfloor \frac{p_{i,l} \cdot M_p}{\rho} \right\rfloor \cdot x_{i,l}$$

$$\geqslant \sum_{(i,l) \in J} \frac{p_{i,l} \cdot M_p}{\rho} \cdot x_{i,l} - |J| \geqslant \frac{M_p}{\rho} \cdot p(x) - k'$$

$$\geqslant \frac{M_p}{\rho} \cdot \rho \cdot (1 + \varepsilon'') - k' = M_p + M_p \cdot \varepsilon'' - k'$$
$$= M_p + \left\lceil \frac{k'}{\varepsilon''} \right\rceil \cdot \varepsilon'' - k' \geqslant M_p. \qquad \square$$

Claim: If $\overline{w}(x) \leqslant M_w$, it holds that $w(x) \leqslant \omega$. Moreover, if $w(x) \leqslant \rho \cdot (1 - \varepsilon')$, we get that $\overline{w}(x) \leqslant M_w$.

Proof: Analogously to the previous claim, suppose that $\overline{w}(x) \leqslant M_w$. For the weight $w(x)$, we get that

$$w(x) = \sum_{(i,l) \in K'} w_{i,l} \cdot x_{i,l} = \frac{\omega}{M_w} \cdot \sum_{(i,l) \in K'} \frac{w_{i,l} \cdot M_w}{\omega} \cdot x_{i,l}$$
$$\leqslant \frac{\omega}{M_w} \cdot \sum_{(i,l) \in K'} \overline{w}_{i,l} \cdot x_{i,l} = \frac{\omega}{M_w} \cdot \overline{w}(x) \leqslant \frac{\omega}{M_w} \cdot M_w = \omega.$$

Now suppose that $w(x) \leqslant \omega \cdot (1 - \varepsilon')$. Again, let $J \subseteq K'$ be the set of all $(i,l) \in K'$ with $x_{i,l} = 1$. We then get the following upper bound on $\overline{w}(x)$:

$$\overline{w}(x) = \sum_{(i,l) \in K'} \overline{w}_{i,l} \cdot x_{i,l} = \sum_{(i,l) \in J} \left\lceil \frac{w_{i,l} \cdot M_w}{\omega} \right\rceil \cdot x_{i,l}$$
$$\leqslant \sum_{(i,l) \in J} \frac{w_{i,l} \cdot M_w}{\omega} \cdot x_{i,l} + |J| \leqslant \frac{M_w}{\omega} \cdot w(x) + k'$$
$$\leqslant \frac{M_w}{\omega} \cdot \omega \cdot (1 - \varepsilon') + k' = M_w - M_w \cdot \varepsilon' + k'$$
$$= M_w - \left\lceil \frac{k'}{\varepsilon'} \right\rceil \cdot \varepsilon' + k' \leqslant M_w. \qquad \square$$

The above two claims show that it is possible to solve the gap-problem by checking if there is a solution $x \in X$ for the scaled instance I'' with $\overline{p}(x) \geqslant M_p$ and $\overline{w}(x) \leqslant M_w$. It remains to show that such a solution can be computed in polynomial time.

Consider the binary tree structure T representing the laminar cardinality constraints of I' (or, equivalently, I'') as introduced in the proof of Theorem 5.15 and let $A_{T'}(\overline{p}, \overline{w})$ denote the minimum number of items in the subtree T' of T that are needed in order to achieve a profit of at least $\overline{p}$ with a weight of at most $\overline{w}$. For a leaf T' of the tree representing some item $(i,l) \in K'$, we get that

$$A_{T'}(\overline{p}, \overline{w}) = \begin{cases} n_{i,l}, & \text{if } \overline{p}_{i,l} \geqslant \overline{p} \text{ and } \overline{w}_{i,l} \leqslant \overline{w}, \\ +\infty, & \text{else.} \end{cases}$$

Similarly, for some subtree T' of T whose root node represents some cardinality constraint $\sum_{i \in I_j} \sum_{l \in \{0,\dots,\bar{l}_i\}} \eta_{i,l} \cdot x_{i,l} \leqslant \mu_j$ and that has two children in the subtrees T_1 and T_2, we get the following formulation for $A_{T'}(\bar{p}, \bar{w})$:

$$A_{T'}(\bar{p}, \bar{w}) = \begin{cases} \mu_{T'}(\bar{p}, \bar{w}), & \text{if } \mu_{T'}(\bar{p}, \bar{w}) \leqslant \mu_j, \\ +\infty, & \text{else,} \end{cases}$$

with

$$\mu_{T'}(\bar{p}, \bar{w}) := \min_{0 \leqslant \bar{p}_1 \leqslant \bar{p}} \min_{0 \leqslant \bar{w}_1 \leqslant \bar{w}} A_{T_1}(\bar{p}_1, \bar{w}_1) + A_{T_2}(\bar{p} - \bar{p}_1, \bar{w} - \bar{w}_1).$$

Hence, by computing the values $A_{T'}(\bar{p}, \bar{w})$ for $\bar{p} \in \{0, \dots, M_p\}$, $\bar{w} \in \{0, \dots, M_w\}$, and for each subtree T' of T in a bottom-up manner and checking if $A_T(M_p, M_w) < +\infty$, we can decide if there is a $x \in X$ with $\bar{p}(x) \geqslant M_p$ and $\bar{w}(x) \leqslant M_w$, which allows us to solve the gap-problem using the above two claims. Hence, the running time of the procedure is bounded by

$$\mathcal{O}\left(M_p^2 \cdot M_w^2 \cdot k'\right) = \mathcal{O}\left(\left(\frac{k'}{\varepsilon''}\right)^2 \cdot \left(\frac{k'}{\varepsilon'}\right)^2 \cdot k'\right) = \mathcal{O}\left(\frac{k^5}{\varepsilon^4} \cdot \log^5 \bar{k}\right)$$

since there are $\mathcal{O}(M_p \cdot M_w)$ table entries for each of the $\mathcal{O}(k')$ subtrees T' of T and it takes $\mathcal{O}(M_p \cdot M_w)$ time to compute each entry in the worst case. Note that the solution $x \in X$ can be computed alongside with the above dynamic programming scheme by keeping track of the variables that led to the minimum in each step. $\qquad \square$

Theorem 5.20:
An ε-approximate pareto frontier $\mathcal{P}(\varepsilon)$ for a given instance of LKP can be computed in $\mathcal{O}\left(\frac{k^5}{\varepsilon^5} \cdot \log^5 \bar{k} \cdot (\log P + \log W)\right)$ time, where P and W denote the maximum profit and weight of a feasible solution, respectively.

Proof: Consider the objective space $\{0, \dots, P\} \times \{0, \dots, W\}$ that represents a superset of all possible combinations of profits and weights of feasible solutions $x \in X$, where P and W are the maximum profit and weight of a feasible solution, respectively. For $\varepsilon'' := \sqrt{1 + \varepsilon} - 1$ as above, we divide this space into rectangles in a way such that, in each dimension, the ratio between the two endpoints of each rectangle is $1 + \varepsilon''$. Obviously, the total number of such rectangles is given by

$$\mathcal{O}\left(\log_{1+\varepsilon''}(P) \cdot \log_{1+\varepsilon''}(W)\right) = \mathcal{O}\left(\frac{\log P \cdot \log W}{(\log(1 + \varepsilon''))^2}\right) = \mathcal{O}\left(\frac{\log P \cdot \log W}{(\varepsilon'')^2}\right)$$
$$= \mathcal{O}\left(\frac{\log P \cdot \log W}{\varepsilon^2}\right),$$

where the second equality follows from the fact that $\frac{1}{\log_a(1+z)} \leqslant \frac{1}{\log_a(2) \cdot z}$ for $z \in (0,1)$. For each corner point (ρ, ω) of these rectangles, we solve the gap-problem in $\mathcal{O}\left(\frac{k^5}{\varepsilon^4} \cdot \log^5 \overline{k}\right)$ time according to Lemma 5.19. If the gap-problem returns some solution $x \in X$, we add the point x to $\mathcal{P}(\varepsilon)$. After each point is considered, the algorithm removes each point in $\mathcal{P}(\varepsilon)$ that is dominated[2] by another point in $\mathcal{P}(\varepsilon)$ within the same time-bound.

It remains to show that the set of points returned by the algorithm forms an ε-approximate pareto frontier. Consider some point $x \in X$ and assume that there was no point $x_P \in \mathcal{P}(\varepsilon)$ with $(1 + \varepsilon) \cdot p(x_P) \geqslant p(x)$ and $w(x_P) \leqslant (1 + \varepsilon) \cdot w(x)$. Let (ρ_0, ω_0) be the (unique) corner point considered by the algorithm that fulfills $\rho_0 \leqslant p(x) < \rho_0 \cdot (1 + \varepsilon'')$ and $\omega_0 \geqslant w(x) > \frac{\omega_0}{1+\varepsilon''}$. Furthermore, let (ρ, ω) be the point that is defined as $\rho := \frac{\rho_0}{1+\varepsilon''}$ and $\omega := \omega_0 \cdot (1 + \varepsilon'')$. The situation is depicted in Figure 5.4.

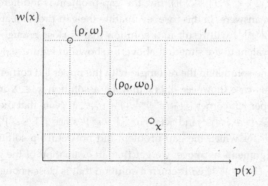

Figure 5.4: The point x that is assumed to be too far away from the ε-approximate pareto frontier and the two pairs (ρ, ω) and (ρ_0, ω_0) considered by the algorithm.

When the gap-problem is called for the point (ρ, ω), it will behave deterministically and return some solution x_P with $p(x_P) \geqslant \rho$ and $w(x_P) \leqslant \omega$ since the point x satisfies

$$p(x) \geqslant \rho_0 = \rho \cdot (1 + \varepsilon'') \text{ and}$$
$$w(x) \leqslant \omega_0 = \frac{\omega}{1+\varepsilon''} = \frac{\omega}{\sqrt{1+\varepsilon}} = (1 - \varepsilon') \cdot \omega.$$

But the returned point x_P furthermore satisfies

$$p(x_P) \geqslant \rho = \frac{\rho_0}{1+\varepsilon''} > \frac{p(x)}{(1+\varepsilon'')^2} = \frac{p(x)}{1+\varepsilon} \text{ and}$$

---

2 A point $x \in X$ is called *dominated* by a point $x' \in X$ if $p(x') \geqslant p(x)$, $w(x') \leqslant w(x)$ and $(p(x), w(x)) \neq (p(x'), w(x'))$.

$$w(x_P) \leqslant \omega = (1 + \varepsilon'') \cdot \omega_0 < (1 + \varepsilon'')^2 \cdot w(x) = (1 + \varepsilon) \cdot w(x).$$

Thus, the point x_P (or some other point in $\mathcal{P}(\varepsilon)$ that dominates x_P) fulfills the property stated in Definition 5.17, which contradicts the assumption. Hence, the set $P(\varepsilon)$ computed by the above algorithm is in fact an ε-approximate pareto frontier.

The number of corner points that need to be considered by the above algorithm can be reduced from $\mathcal{O}\left(\frac{\log P \cdot \log W}{\varepsilon^2}\right)$ to $\mathcal{O}\left(\frac{\log P + \log W}{\varepsilon}\right)$ by a more sophisticated approach: Let j be an integer that is initially set to zero. Starting with $i := 0$ and increasing i by one in each iteration, we consider the pairs $((1 + \varepsilon'')^i, (1 + \varepsilon'')^j)$ and add the solution returned by the gap-problem to $\mathcal{P}(\varepsilon)$ until we reach a point $((1 + \varepsilon'')^i, (1 + \varepsilon'')^j)$ for which the gap-problem does not return a solution. We differentiate between the following two cases:

Case 1: There is a solution $x \in X$ that is contained in the rectangle with the upper left corner $((1 + \varepsilon'')^i, (1 + \varepsilon'')^j)$, but the gap-problem is indifferent between its two possible answers. In this case, evaluating the gap-problem at the point $((1 + \varepsilon'')^{i-1}, (1 + \varepsilon'')^{j+1})$ will result in a solution that is close enough to the point x (this is equivalent to the situation above, as shown in Figure 5.4).

Case 2: There is no solution in the rectangle with the upper left corner $((1 + \varepsilon'')^i, (1 + \varepsilon'')^j)$. In this case, however, there may be a solution $x \in X$ in the rectangle with the upper left corner $((1 + \varepsilon'')^{i-1}, (1 + \varepsilon'')^{j+1})$. Note that there is a point x' with $p(x') \geqslant (1 + \varepsilon'')^{i-1}$ and $w(x') \leqslant (1 + \varepsilon'')^j$ since $((1 + \varepsilon'')^i, (1 + \varepsilon'')^j)$ was the first pair for which the gap-problem did not return a solution. Following the same arguments as above, we get that an evaluation of the gap-problem at $((1 + \varepsilon'')^{i-2}, (1 + \varepsilon'')^{j+1})$ will return a solution that is close enough to x.

The above arguments show that we can proceed by setting $j := j + 1$ and $i := i - 2$ without "missing" a point that needs to be added to $\mathcal{P}(\varepsilon)$. Note that this procedure only considers a linear number $\mathcal{O}\left(\frac{\log P + \log W}{\varepsilon}\right)$ of points, which results in an improved running time. $\qquad\square$

Corollary 5.21:

An ε-approximate pareto frontier $\mathcal{P}(\varepsilon)$ for BCMCFP$_\mathbb{N}$ on extension-parallel graphs can be computed in $\mathcal{O}\left(\frac{m^5}{\varepsilon^5} \cdot \log^5 U \cdot (\log mCU + \log B)\right) = \mathcal{O}\left(\frac{m^5}{\varepsilon^5} \cdot \log^6 M\right)$ time. $\qquad\square$

Theorem 5.20 also yields a bicriteria FPTAS for LKP, which can be seen as follows: Let x^* denote an optimal solution to LKP, which consequently fulfills $w(x^*) \leqslant W$. According to the definition of $\mathcal{P}(\varepsilon)$, there is a point $x_P \in P(\varepsilon)$ such that

$$p(x_P) \geqslant \frac{1}{1 + \varepsilon} \cdot p(x^*) \geqslant (1 - \varepsilon) \cdot p(x^*) \text{ and}$$
$$w(x_P) \leqslant (1 + \varepsilon) \cdot w(x^*).$$

By computing an ε-approximate pareto frontier of LKP and searching for the point x_P with the largest profit $p(x_P)$ fulfilling $w(x_P) \leqslant (1 + \varepsilon) \cdot W$ among all points in $\mathcal{P}(\varepsilon)$, we obtain a solution of LKP that fulfills the above properties. This yields the following corollary:

Corollary 5.22:
There is a bicriteria FPTAS for LKP running in $\mathcal{O}\left(\frac{k^5}{\varepsilon^5} \cdot \log^5 \overline{k} \cdot (\log P + \log W)\right)$ time. $\qquad\square$

Corollary 5.23:
There is a bicriteria FTPAS for BCMCFP$_\mathbb{N}$ on extension-parallel graphs running in $\mathcal{O}\left(\frac{m^5}{\varepsilon^5} \cdot \log^5 U \cdot (\log mCU + \log B)\right) = \mathcal{O}\left(\frac{m^5}{\varepsilon^5} \cdot \log^6 M\right)$ time. $\qquad\square$

5.4.3 A Polynomial-Time Solvable Special Case

For the case that we are dealing with the traditional bounded knapsack problem, we are able to solve the problem for bounded weights optimally in polynomial time. Note that this is trivial for the traditional 0-1-knapsack problem, but not for the bounded knapsack problem.

Theorem 5.24:
LKP is solvable in polynomial time $\mathcal{O}(k^3 \cdot \overline{w}^3)$ if the weights w_i for $i \in K$ are polynomially bounded by some value $\overline{w}$ and if $h = 0$.

Proof: Let x^* be an optimal solution of the problem and let $\overline{w} := \max_{i \in K} w_i$. Clearly, we may assume that $w_i > 0$ for each $i \in K$ since we may else set $x_i := k_i$ in any solution x and remove the item type from the instance. Furthermore, we assume that the item types are sorted in non-increasing order by their profit-per-weight ratio, i.e., $\frac{p_i}{w_i} \geqslant \frac{p_j}{w_j}$ for $i < j$, which can be established in $\mathcal{O}(k \log k)$ time.

Now suppose that there are indices $i, j \in K$ with $i < j$ such that $x_i^* \leqslant k_i - \overline{w}$ and $x_j^* \geqslant \overline{w}$. By replacing w_i items of type j with w_j items of type i, the total weight of the knapsack clearly does not change. The total profit changes by $p_i \cdot w_j - p_j \cdot w_i$. Due to the ordering of the item types, we have

$$\frac{p_i}{w_i} \geqslant \frac{p_j}{w_j} \quad \Longleftrightarrow \quad p_i \cdot w_j \geqslant p_j \cdot w_i,$$

so the new solution is optimal, again.

Hence, we may assume without loss of generality that $x_i^* > k_i - \overline{w}$ or $x_j^* < \overline{w}$ for each $i, j \in K$ with $i < j$. In the following, we differentiate between the following two cases:

Case 1: $x_j^* < \overline{w}$ **for all** $j \in K$: In this case, each item type is packed at most $\overline{w} - 1$ times in x^*, which is a polynomially bounded number. We are thus able to transform the bounded knapsack problem into a traditional 0-1-knapsack problem in polynomial time by introducing $\overline{w} - 1$ copies of each item type. Furthermore, we may restrict the maximum weight to $W' := \min\{W, \sum_{i \in K}(\overline{w} - 1) \cdot w_i\} = \mathcal{O}(k \cdot \overline{w}^2)$. By using the dynamic programming approach as described in (Kellerer et al., 2004), which solves the traditional knapsack problem with $\widehat{k}$ items and a knapsack capacity of $\widehat{W}$ in $\mathcal{O}(\widehat{k} \cdot \widehat{W})$ time, we can solve this special case in polynomial time $\mathcal{O}((k \cdot (\overline{w} - 1)) \cdot (k \cdot \overline{w}^2)) = \mathcal{O}(k^2 \cdot \overline{w}^3)$.

Case 2: $x_h^* \geqslant \overline{w}$ **for some** $h \in K$: Let h be the largest index with $x_h^* \geqslant \overline{w}$. Note that $x_i^* > k_i - \overline{w}$ for all $i < h$ without loss of generality since we can otherwise replace w_i items of type h with w_h items of type i as above. Thus, we get that $x_i^* > k_i - \overline{w}$ for all $i < h$ and, by the definition of h, that $x_j^* < \overline{w}$ for all $j > h$. Hence, except for item type h, we must only check a polynomially bounded number of possibilities of how to pack each item type. Note that we can discard the current distribution scheme if $W_0 := W - \sum_{i < h}(k_i - \overline{w} + 1) \cdot w_i < 0$ since it will not be possible to fulfill $x_i^* > k_i - \overline{w}$ for $i < h$ without exceeding the knapsack capacity. Else, we set $W' := \min\{W_0, \sum_{i \in K \setminus \{h\}}(\overline{w} - 1) \cdot w_i\}$ and use the dynamic programming scheme as described in the first case but without item type h, which can be done in $\mathcal{O}(k^2 \cdot \overline{w}^3)$ time. Since we only packed item types in $K \setminus \{h\}$ so far, we afterwards need to try out each combination of a packing of the item types in $K \setminus \{h\}$ with some maximum weight $W_1 \leqslant W'$ and the maximum amount of item type h that neither exceeds the knapsack capacity W nor k_h. This can be done in $\mathcal{O}(W') = \mathcal{O}(k \cdot \overline{w}^2)$ time using the table created in the dynamic programming scheme before.

Hence, for each possible position for h and the case that such an index h does not exist, we are able to solve the problem in polynomial time $\mathcal{O}(k^2 \cdot \overline{w}^3)$. Since we do not know h in advance, we need to try out each of the $k + 1$ possible cases and maximize over the resulting total profit. This yields a total running time of $\mathcal{O}(k^3 \cdot \overline{w}^3)$, which shows the claim. $\square$

We are now able to apply the result from Theorem 5.24 to the case of BCMCFP_N with upgrade costs that are polynomially bounded by some value $\overline{b}$ since the total upgrade cost per unit of flow of each s-t-path is then bounded by $n \cdot \overline{b}$. The restriction that there is no cardinality constraint corresponds to the restriction that the paths in the corresponding instance of BCMCFP_N have no edge in common. Hence, we immediately get the following result:

Corollary 5.25:
$BCMCFP_N$ is solvable on edge-disjoint s-t-paths in polynomial time $\mathcal{O}(n^3 \cdot m^3 \cdot \overline{b}^3)$ if the upgrade cost b_e for $e \in E$ are polynomially bounded by some value $\overline{b}$. $\quad\square$

Note that the above argumentation yields a pseudo-polynomial-time algorithm for $BCMCFP_N$ on such graphs in case that the upgrade costs are not polynomially bounded. In contrast to the algorithm obtained in Theorem 5.5, the running time of this algorithm only depends on the maximum upgrade costs and the number of edges and nodes and may be useful in case that the edge capacities are large.

Moreover, note that the situation of parallel s-t-paths, for which Corollary 5.25 is applicable, includes the graph that was used at the beginning of this chapter in the proof of Theorem 5.3, in which weak $\mathcal{NP}$-completeness of the problem was shown. In fact, the reduction from SUBSETSUM that was used in the corresponding proof relies on the fact that the upgrade costs are *not* polynomially bounded. This closes the discussion of the problem $BCMCFP_N$.

5.5 Binary Case

We conclude our considerations with the second variant of how to calculate the upgrade costs, i.e., we assume that $b(x) = b_B(x) = \sum_{e \in E} b_e \cdot u_e \cdot sgn(x_e)$. Whenever flow is sent through some edge $e \in E$, it must be upgraded to its maximum capacity u_e, which yields upgrade costs of $b_e \cdot u_e$. This problem can be seen as a generalization of the MAXFIXEDCOSTFLOW-problem, in which the aim is to determine a maximum flow while fixed costs are incurred by the usage of edges that must fulfill a given budget constraint (cf. (Garey and Johnson, 1979, Problem ND32) and (Krumke and Schwarz, 1998)). As we will see, we are able to adapt several results from the previous sections to the case of $BCMCFP_B$.

Obviously, the problem $BCMCFP_B$ coincides with $BCMCFP_N$ in case that the capacities are zero or one. Nevertheless, in contrast to the case of $BCMCFP_N$, the problem $BCMCFP_B$ turns out to be strongly $\mathcal{NP}$-complete to solve if we allow larger capacities, as we will see in the following theorem:

Theorem 5.26:
$BCMCFP_B$ is strongly $\mathcal{NP}$-complete to solve even on bipartite graphs.

Proof: Clearly, the decision version of $BCMCFP_B$ lies in the class $\mathcal{NP}$ since we can verify any solution (which has a polynomially bounded encoding length) in polynomial time. For the reduction to prove $\mathcal{NP}$-hardness, we use the EXACTCOVERBY3SETS-problem, which is known to be strongly $\mathcal{NP}$-complete to solve (Garey and Johnson, 1979, Problem SP2):

INSTANCE: Set X with 3q elements and a collection $\mathcal{C} = \{C_1, \ldots, C_k\}$ of 3-element subsets of X.

QUESTION: Does there exist a subcollection $\mathcal{C}' \subseteq \mathcal{C}$ such that every element $j \in X$ is contained in exactly one of the subsets in $\mathcal{C}'$?

Given an instance of EXACTCOVERBY3SETS, we construct an instance of BCMCFP$_B$ as follows:

We introduce a source s and a sink t as well as a node v_i for each $C_i \in \mathcal{C}$ and a node v_j' for each $j \in X$. For each subset $C_i \in \mathcal{C}$, we insert an edge with cost 0, upgrade cost 1, and capacity 3 that connects s to v_i. Furthermore, we introduce an edge with cost 0, upgrade cost 0, and capacity 1 between v_i and v_j', if $j \in C_i$, and an edge between each node v_j' and the sink t with cost -1, upgrade cost 0, and capacity 1. The budget is set to $B := 3q$. The resulting network for $X = \{1, \ldots, 9\}$ and $\mathcal{C} = \{\{1,2,4\}, \{2,3,4\}, \{3,5,8\}, \{4,6,7\}, \{6,7,9\}\}$ is shown in Figure 5.5.

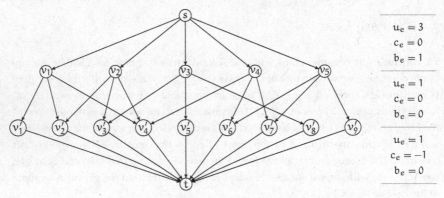

Figure 5.5: The resulting network for a given instance of EXACTCOVERBY3SETS with $X = \{1, \ldots, 9\}$ and $\mathcal{C} = \{\{1,2,4\}, \{2,3,4\}, \{3,5,8\}, \{4,6,7\}, \{6,7,9\}\}$. On the right hand side, the capacities, costs, and upgrade costs of the edges in each level of the graph are depicted.

We claim that there is a flow x with costs $c(x) \leqslant -3q$ if and only if the given instance of EXACTCOVERBY3SETS is a YES-instance.

Assume that there is a flow x with $c(x) \leqslant -3q$. Since the capacity of each edge that leads to the sink is 1, the costs are -1, and there are 3q such edges in total, each of these edges must carry exactly one unit of flow. Furthermore, since the budget is set to 3q, only q of the k edges leaving the source can be upgraded and must carry 3 units of flow each. This flow, which arrives at some of the nodes v_i, must be distributed to each of the nodes v_j' in a way such that every v_j' receives one unit of flow from exactly one of the used nodes v_i. By identifying each node v_i with positive inflow (and outflow) with its corresponding set C_i, we thus get a solution of EXACTCOVERBY3SETS.

Conversely, assume that there is a solution $\mathcal{C}' \subseteq \mathcal{C}$ of EXACTCOVERBY3SETS. We identify each set $C_i \in \mathcal{C}'$ with its corresponding node v_i in the same way as before and upgrade the edge leading to v_i at a total upgrade costs of $3q = B$. Since the sets $C \in \mathcal{C}'$ are disjoint and their union is X, we can send one unit of flow to each v_j' by sending 3 units of flow to each node v_i and distributing this flow to its three adjacent nodes v_j'. This way, we achieve a feasible flow with cost $-3q$, which proves the claim. $\square$

In addition to the result obtained in the previous theorem, we can immediately conclude from Theorem 5.4 that BCMCFP$_\mathbb{B}$ is weakly $\mathcal{NP}$-hard to solve and to approximate on bipartite series-parallel graphs since we used capacities $u_e = 1$ for each edge $e \in E$ in the proof of Theorem 5.4 for which case BCMCFP$_\mathbb{B}$ and BCMCFP$_\mathbb{N}$ obviously coincide.

Corollary 5.27:
BCMCFP$_\mathbb{B}$ is weakly $\mathcal{NP}$-hard (to solve and) to approximate, even on bipartite series-parallel graphs. $\square$

Similarly, we immediately get that BCMCFP$_\mathbb{B}$ is weakly $\mathcal{NP}$-complete to solve on parallel edges:

Corollary 5.28:
BCMCFP$_\mathbb{B}$ is weakly $\mathcal{NP}$-complete to solve, even when $u_e = 1$ for each $e \in E$ and the graph consists of parallel edges only. $\square$

Moreover, it can be easily seen that the pseudo-polynomial-time algorithm for the problem BCMCFP$_\mathbb{N}$ as introduced in Theorem 5.5 can be modified easily for the case of BCMCFP$_\mathbb{B}$:

Theorem 5.29:
BCMCFP$_\mathbb{B}$ can be solved in pseudo-polynomial time $\mathcal{O}(nmCUB^2(n^2C + m^2U))$ on series-parallel graphs.

Proof: The algorithm works analogously to the one introduced in Theorem 5.5. Solely the case that a considered subgraph G' that corresponds to a node in the decomposition tree of G consists of a single edge e must now incorporate the "all-or-nothing" aspect of BCMCFP$_\mathbb{B}$. As in Theorem 5.5, we set $A_{G'}(c, b, f) := \infty$ if the flow value f is demanded to be positive and larger than u_e, the costs c_e exceed the bound c, or the upgrade of the edge would exceed the budget b. The minimum costs are, thus, given by the following expression:

$$A_{G'}(c,b,f) := \begin{cases} f \cdot c_e, & \text{if } (c_e \leqslant c \wedge f \leqslant u_e \wedge u_e \cdot b_e \leqslant b) \vee f = 0, \\ +\infty, & \text{else.} \end{cases} \tag{5.6}$$

Hence, equation (5.6) together with equations (5.2) and (5.3) from Theorem 5.5 can be used to compute the costs of a minimum cost flow for the given instance of BCMCFP_B. By proceeding as in the proof of Theorem 5.5, we are then able to compute a budget-constrained minimum cost flow within the claimed running time. $\qquad\square$

Corollary 5.30:
BCMCFP_B can be solved in pseudo-polynomial time $\mathcal{O}(m^3 U^2 B^2)$ time on extension-parallel graphs.

Consider the *binary bounded knapsack problem with laminar cardinality constraints* (BLKP) that is the variant of LKP in which the weight of an item type i is no longer proportional to the number of packed items of this type but is $w_i \cdot k_i$ if at least one item of type i is packed into the knapsack and zero else. It is easy to see that, on extension-parallel graphs, there is the same analogy between BCMCFP_B and BLKP as between BCMCFP_N and LKP.

Theorem 5.31:
There is an FPTAS for BLKP that runs in $\mathcal{O}\left(k \cdot \overline{\mu}^6 \cdot \frac{1}{\varepsilon^2}\right)$ time if the maximum number μ of items in an optimal solution is polynomially bounded.

Proof: The proof is similar to the one of Theorem 5.15. Again consider the binary tree T that represents the subset relation between the sets I_j for $j \in \{1, \ldots, h\}$. Let $A_{T'}(\mu, p)$ now denote the minimum weight that is needed in order to create a total profit of at least p with item types that are contained in the subtree T' of T while using *exactly* μ such items. For the case that T' is a leaf of T, i.e., T' corresponds to a single item type $i \in K$, we now have

$$A_{T'}(\mu, p) := \begin{cases} 0, & \text{if } \mu = 0 \wedge p = 0, \\ w_i \cdot k_i, & \text{if } \left\lceil \frac{p}{p_i} \right\rceil \leqslant \mu \leqslant k_i \\ \infty, & \text{else.} \end{cases}$$

The rest of the proof remains analogously to the case of LKP. However, since a 2-approximation algorithm for BLKP is not known, the value of M in the proof of Theorem 5.15 is changed to $M := \frac{\varepsilon \cdot p_{max}}{\overline{\mu}}$, where $p_{max} := \max\{p_i : i \in K\}$ and $\overline{\mu} :=$

$\sum_{i \in K} k_i$ as before. The solution that is obtained by the described algorithm using the profits $\tilde{p}_i := \lfloor \frac{p_i}{M} \rfloor$ for $i \in K$ then fulfills

$$p(x) = \sum_{i \in K} p_i \cdot x_i \geqslant \sum_{i \in K} M \cdot \left\lfloor \frac{p_i}{M} \right\rfloor \cdot x_i \geqslant \sum_{i \in K} M \cdot \left\lfloor \frac{p_i}{M} \right\rfloor \cdot x_i^*$$

$$\geqslant \sum_{i \in K} M \cdot \left(\frac{p_i}{M} - 1 \right) \cdot x_i^* = \sum_{i \in K} (p_i - M) \cdot x_i^* = p(x^*) - M \cdot \sum_{i \in K} x_i^*$$

$$\geqslant p(x^*) - M \cdot \overline{\mu} = p(x^*) - \varepsilon \cdot p_{max}$$

$$\geqslant p(x^*) - \varepsilon \cdot p(x^*) = (1 - \varepsilon) \cdot p(x^*).$$

Moreover, each solution $\tilde{x}$ of an instance of LKP with the profits $\tilde{p}_i$ satisfies

$$\tilde{p}(\tilde{x}) := \sum_{i \in K} \tilde{p}_i \cdot \tilde{x}_i = \sum_{i \in K} \left\lfloor \frac{p_i}{M} \right\rfloor \cdot \tilde{x}_i \leqslant \frac{1}{M} \sum_{i \in K} p_i \cdot \tilde{x}_i \leqslant \frac{1}{M} \cdot p(x^*)$$

$$\leqslant \frac{1}{M} \cdot p_{max} \cdot \overline{\mu} = \frac{\overline{\mu}^2}{\varepsilon},$$

so the maximum possible total profit $\tilde{P}$ is bounded by $\frac{\overline{\mu}^2}{\varepsilon}$. Since the algorithm that was described in Theorem 5.15 runs in $\mathcal{O}\left(k \cdot \overline{\mu}^2 \cdot \tilde{P}^2 \right)$ time, this yields a running time of $\mathcal{O}\left(k \cdot \overline{\mu}^6 \cdot \frac{1}{\varepsilon^2} \right)$ for the overall procedure. □

Corollary 5.32:
There is an FPTAS for BCMCFP$_\mathbb{B}$ on extension-parallel graphs with a running time of $\mathcal{O}\left(m \cdot F^6 \cdot \frac{1}{\varepsilon^2} \right)$ if the maximum flow value F is polynomially bounded. □

5.6 Conclusion

We studied two discrete natural extensions of the budget-constrained minimum cost flow problem that extend the possible applications of the continuous case considered in Chapter 4. We saw that these problems are equivalent to a minimum cost flow problem in which the capacities of the edges must be upgraded by a sufficiently large amount in order to send flow through them. For the first variant BCMCFP$_\mathbb{N}$, which requires integral upgrades, we showed $\mathcal{NP}$-hardness of solving and approximating the problem on restricted graph classes. However, we found a pseudo-polynomial-time algorithm for the problem on series-parallel graphs and observed an analogy between the budget-constrained minimum cost flow problem on extension-parallel graphs and an extension of the bounded knapsack problem by cardinality constraints that fulfill a special laminarity property. This observation allowed us to derive a PTAS for the general case and an FPTAS for the case of bounded edge capacities as well as a polynomial-time algorithm to determine an ε-approximate pareto frontier, which in turn implied a bicriteria FPTAS for the problem. Moreover, we identified a polynomial-time solvable special case for the problem on edge-disjoint s-t-paths with bounded upgrade costs. Finally, we studied the case that edges must be upgraded up to their full capacity (BCMCFP$_\mathbb{B}$), showed strong $\mathcal{NP}$-completeness, and adapted the pseudo-polynomial-time algorithm and the FPTAS for BCMCFP$_\mathbb{N}$ variant to this binary variant. A complete overview of the results is given in Table 5.1 and Table 5.2.

Acyclic Graphs	Series-Parallel Graphs	Extension-Parallel Graphs
	Theorem 5.5: Solvable in $\mathcal{O}(nmCUB^2 \cdot (n^2C + m^2U))$ time	Corollary 5.6: Solvable in $\mathcal{O}(m^3U^2B^2)$ time
$\longleftarrow$	$\longleftarrow$	Theorem 5.3: weakly $\mathcal{NP}$-complete to solve
$\longleftarrow$	Theorem 5.4: $\mathcal{NP}$-hard to approximate	Corollary 5.12: 2-approximation in $\mathcal{O}(m^2 \log^2 m)$ time
		Corollary 5.14: PTAS in $\mathcal{O}((\frac{m}{\varepsilon})^{\lceil \frac{1}{\varepsilon} \rceil - 2} \cdot (\frac{m}{\varepsilon} + m^2 \log^2 m))$ time
		Corollary 5.16: FPTAS in $\mathcal{O}\left(m \cdot F^4 \cdot \frac{1}{\varepsilon^2}\right)$ time
		Corollary 5.23: Bicriteria FPTAS in $\mathcal{O}\left(\frac{m^5}{\varepsilon^5} \cdot \log^6 M\right)$ time
		Corollary 5.25: Solvable in $\mathcal{O}(n^3 \cdot m^3 \cdot \overline{b}^3)$ time on node-disjoint s-t-paths

Table 5.1: The summarized results for BCMCFP$_\mathbb{N}$ in Chapter 5. Implied results are denotes with gray arrows.

The introduced models raise several questions for future research. Especially for the case of BCMCFP$_\mathbb{N}$, it is unclear if there is a pseudo-polynomial-time algorithm for general graphs or if the problem becomes strongly $\mathcal{NP}$-complete to solve in this setting. Furthermore, approximation algorithms for more general cases would be interesting to investigate. Finally, an extension by the possibility to upgrade costs and/or to

Acyclic Graphs	Series-Parallel Graphs	Extension-Parallel Graphs
	Theorem 5.29: Solvable in $\mathcal{O}(nmCUB^2 \cdot (n^2C + m^2U))$ time	**Corollary 5.30:** Solvable in $\mathcal{O}(m^3U^2B^2)$ time
Theorem 5.26: Strongly $\mathcal{NP}$-complete to solve	⟵	**Corollary 5.28:** Weakly $\mathcal{NP}$-complete to solve
⟵	**Corollary 5.27:** $\mathcal{NP}$-hard to approximate	**Corollary 5.32:** FPTAS in $\mathcal{O}\left(m \cdot F^6 \cdot \frac{1}{\epsilon^2}\right)$ time

Table 5.2: The summarized results for BCMCFP$_\mathbb{B}$ in Chapter 5. Implied results are denoted with gray arrows.

upgrade all the edges that are incident to the same node at once are interesting topics for further research.

6 | Generalized Processing Networks

We turn our considerations to a generalization of the maximum flow problem in which each edge $e = (v,w) \in E$ is assigned with a so called *flow ratio* $\alpha_e \in [0,1]$ that imposes an upper bound on the fraction of the total outgoing flow at v that may be routed through the edge e. This model embodies a generalization of the maximum flow problem in *processing networks* (Koene, 1982), in which the corresponding flow ratios specify the *exact* fraction of flow rather than only an upper bound. We show that a flow decomposition similar to the one for traditional network flows is possible and can be computed in strongly polynomial time. Moreover, we prove that the problem is at least as hard to solve as any packing LP but that there also exists a fully polynomial-time approximation scheme for the maximum flow problem in these generalized processing networks if the underlying graph is acyclic. For the case of series-parallel graphs, we provide two exact algorithms with strongly polynomial running time. Finally, we study the case of *integral* flows and show that the problem becomes $\mathcal{NP}$-hard to solve and approximate in this case.

This chapter is based on joint work with Sven O. Krumke and Clemens Thielen (Holzhauser et al., 2016c).

6.1 Introduction

Traditional flows in networks that were introduced in Section 2.4 and extended in the previous chapters embody a useful tool to model the transshipment of commodities from nodes with supply to nodes with demand. However, in order to model advanced issues such as the production of goods in a manufacturing process, the considered network flow problems are not powerful enough since they lack the possibility to model the splitting of flow at nodes by specific ratios. *Processing networks* (cf. (Koene, 1982)) generalize traditional flow problems by the introduction of *processing nodes* that involve additional *flow ratios* $\alpha_e \in [0,1]$ for their outgoing edges e. The flow on such an outgoing edge e is required to equal a fraction α_e of the total flow on the outgoing edges of the processing node. In order to maintain flow conservation, these flow ratios of all outgoing edges of each processing node are required to sum up to one.

In this chapter, we investigate a generalization of processing networks in which a flow ratio $\alpha_e \in [0, 1]$ is assigned to *every* edge e. The flow ratios are required to sum up to *at least* one at every node with outgoing edges and only impose an *upper bound* on the ratio of flow on the corresponding edges. Clearly, this extended model subsumes both the maximum flow problem in processing networks and the maximum flow problem in traditional networks but also allows to model more advanced situations. We provide several structural results about flows in such networks and present both approximation and exact algorithms for the maximum flow problem in several special cases of these networks.

The possible applications of our model are manifold. The most natural one is the modeling of distillation processes (e.g., in refineries), in which raw materials are split into intermediate and end products in specific ratios. However, in contrast to traditional processing networks, we are now able to model possible variations in these ratios that are only bounded by specific technical limitations. Similarly, by inverting the direction of each edge, we can model manufacturing processes of goods in which the composition ratios of the basic commodities may vary up to specific upper bounds.

6.1.1 Previous Work

Research on the topic of processing networks has a long history under several different names. To the best of our knowledge, first work was done by Schaefer (1978) who introduced the maximum flow problem in processing networks and a first algorithm for the problem. In particular, he considered the case that there are two kind of nodes: ordinary nodes as in traditional network flow problems and special nodes for which each of the outgoing edges has an assigned value $\alpha_e \in (0, 1)$ that determines the fraction of flow that is routed through the corresponding edge e. In order to maintain flow conservation, these values are required to sum up to one at each special node. Schaefer presented a (super-polynomial-time) algorithm that generalizes the augmenting path algorithm for the traditional maximum flow problem (cf. (Ahuja et al., 1993)). However, the author refrained from giving an exact running time analysis and a complete description on how to handle several special cases that might occur in the course of his algorithm. In the 1980s, Koene (1980) considered the maximum flow problem in a processing network where the only special node (called *processing node*) coincides with the source node s of the network. For this special case, he presented a polynomial-time exact algorithm.

In his PhD-thesis, Koene (1982) later generalized the problem in three ways: Besides the introduction of a third kind of nodes (representing so called *blending processes* that assign proportionality values to the *incoming* edges of a node) and the introduction of *gains* on edges similar to the generalized flow problem (cf. Section 2.4), he considered the more general *minimum cost flow* variant of the problem. He showed that every linear program can be transformed into an instance of this minimum cost flow problem and developed a customized variant of the simplex method in order to solve the problem. This simplex method was later improved by Chen and Engquist (1986) and Chang et al. (1989).

Many years later, in 2003, research on processing networks was revived by Fang and Qi (2003) under the name *manufacturing network flows* in which they derived the algebraic foundations for a network simplex method. In the following years, Lu et al. (2006), Lu et al. (2009), Venkateshan et al. (2008), and Wang and Lin (2009) extended this foundation, partially under the name *minimum distribution cost flow problem*, with the introduction of explicit graph operations that are used in a network simplex algorithm. Moreover, Wang and Lin (2009) showed that the maximum flow problem in a processing network with both processing and blending nodes is at least as hard as the maximum generalized flow problem as introduced in Section 2.4.

The maximum flow variant of the problem was again investigated by Sheu et al. (2006) and Huang (2011). In the former paper, the authors provide a similar algorithm to the very early procedure introduced by Schaefer (1978) with super-exponential running time but neither give a proof of correctness nor handle every special case that may occur. In Huang (2011), the author presents a network simplex method for the problem without processing nodes in combination with computational results.

The case that the corresponding factors do not sum up to one at some nodes was considered in Lu et al. (2006). However, the authors do not assume flow conservation to hold at these nodes and are, thus, able to define preprocessing procedures in order to remove such nodes. To the best of our knowledge, the more general case that is considered in this chapter, in which these factors only provide *upper bounds* on the flow while flow conservation is maintained at each node has not been investigated so far.

6.1.2 Chapter Outline

After defining the *maximum flow problem in generalized processing networks* and the necessary notation in Section 6.2, we show in Section 6.3 that there is a flow decomposition theorem similar to the one for traditional flows (cf. (Ahuja et al., 1993)) and that such a flow decomposition can be computed more efficiently in the case of acyclic graphs. In Section 6.4, we consider the complexity and approximability of the problem. In particular, we show that the problem of finding a maximum flow in a generalized processing network is solvable in weakly polynomial-time on the one hand, but at least as hard to solve as any packing LP on the other hand. Moreover, we present an FPTAS for the problem on acyclic graphs that is based on the generalized packing framework introduced in Section 3.3. To the best of our knowledge, this comprises the first approximation algorithm for the maximum flow problem in processing networks. In Section 6.5, we turn our focus to the case of series-parallel graphs and present two different approaches on how to solve the problem exactly in strongly polynomial time. The first of these approaches is an analogue to the augmenting path algorithm for the traditional maximum flow problem (cf. (Ahuja et al., 1993)) and achieves a running time of $\mathcal{O}(m^2)$ while the second approach exhaustedly uses the inherent structure of series-parallel graphs in order to repeatedly shrink series-parallel subcomponents into single edges, which results in an algorithm with an improved running time of $\mathcal{O}(m \cdot (n + \log m))$. Finally, in Section 6.6, we briefly investigate the case of flows that are required to be integral on every edge. As it turns out, the problem with integral flows becomes strongly $\mathcal{NP}$-complete to solve and to approximate even on bipartite acyclic graphs and weakly $\mathcal{NP}$-complete to solve and to approximate on series-parallel graphs. An overview of the results of this chapter is given in Table 6.1 and Table 6.2 on page 152.

6.2 Preliminaries

We start by defining the maximum flow problem in a directed graph $G = (V, E)$ with *edge capacities* $u_e \in \mathbb{N}$ and *flow ratios* $\alpha_e \in (0, 1]$ on the edges $e \in E$. Let $s \in V$ and $t \in V$ denote a distinguished *source* and *sink* of the network, respectively.

Definition 6.1 (Flow, flow value, maximum flow, static/dynamic capacity constraint):
A function $x \colon E \to \mathbb{R}_{\geqslant 0}$ is called a *feasible flow in a generalized processing network* or just *flow* if $\operatorname{excess}_x(v) := \sum_{e \in \delta^-(v)} x_e - \sum_{e \in \delta^+(v)} x_e = 0$ for each $v \in V \setminus \{s, t\}$ and both $x_e := x(e) \leqslant u_e$ (called the *static capacity constraint for e*) and $x_e \leqslant \alpha_e \cdot \sum_{e' \in \delta^+(v)} x_{e'}$ (called the *dynamic capacity constraint for e*) for each $e = (v, w) \in E$. The *flow value* of a flow x is given by $\operatorname{val}(x) := \operatorname{excess}_x(t)$. A flow x of maximum flow value is called a *maximum flow in a generalized processing network* or just *maximum flow*. $\lhd$

The above definition allows us to define the *maximum flow problem in a generalized processing network*:

Definition 6.2 (Maximum flow problem in a generalized processing network (MFGPN)):

INSTANCE: A directed graph $G = (V, E)$ with source $s \in V$, sink $t \in V$, capacities $u_e \in \mathbb{N}$, and flow ratios $\alpha_e \in (0, 1]$ on the edges $e \in E$ such that $\sum_{e \in \delta^+(v)} \alpha_e \geqslant 1$ for each $v \in V$ with $\delta^+(v) \neq \emptyset$.

TASK: Determine a maximum flow in G. ◁

Note that we have required that $\sum_{e \in \delta^+(v)} \alpha_e \geqslant 1$ for each $v \in V$ with $\delta^+(v) \neq \emptyset$ in Definition 6.2. However, this does not yield any restriction since flow conservation holds at nodes with $\sum_{e \in \delta^+(v)} \alpha_e \in (0, 1)$ only if the flow on the outgoing edges is zero. Consequently, we can find and remove such nodes in a preprocessing step in $\mathcal{O}(n + m)$ time. This fact is held down in the following assumption:

Assumption 6.3: For every node $v \in V$ with $\delta^+(v) \neq \emptyset$, the flow ratios of its outgoing edges fulfill $\sum_{e \in \delta^+(v)} \alpha_e \geqslant 1$. ◁

In addition to Assumption 6.3, we make the following assumptions on the structure of the underlying graph:

Assumption 6.4: For every node $v \in V \setminus \{s, t\}$, it holds that $\delta^+(v) \neq \emptyset$ and $\delta^-(v) \neq \emptyset$. ◁

Assumption 6.5: For every node $v \in V \setminus \{s, t\}$, there is at least one directed path from s to v or from v to t. ◁

Assumption 6.4 does not impose any restriction on the underlying model since the inflow and outflow of every node $v \in V \setminus \{s, t\}$ with $\delta^+(v) = \emptyset$ or $\delta^-(v) = \emptyset$ must equal zero due to flow conservation at v, which implies that the incident edges can be deleted in a preprocessing step. Similarly, Assumption 6.5 yields no restriction since the corresponding connected components do not contribute to the flow value in any flow and can be deleted as well. Note that, for any instance of MFGPN, both assumptions can be established in $\mathcal{O}(n + m)$ time by performing a depth-first search and repeatedly deleting single nodes and edges. The resulting graph is connected, such that we can assume that $n \in \mathcal{O}(m)$ in the following.

Using the above definitions, we can formulate the maximum flow problem in a generalized processing network as a linear program as follows:

$$\max \sum_{e \in \delta^-(t)} x_e - \sum_{e \in \delta^+(t)} x_e \tag{6.1a}$$

$$\text{s.t.} \sum_{e \in \delta^-(v)} x_e - \sum_{e \in \delta^+(v)} x_e = 0 \qquad \text{for all } v \in V \setminus \{s, t\}, \tag{6.1b}$$

$$x_e \leqslant \alpha_e \cdot \sum_{e' \in \delta^+(v)} x_{e'} \qquad\qquad \text{for all } e = (v, w) \in E, \qquad (6.1c)$$

$$0 \leqslant x_e \leqslant u_e \qquad\qquad \text{for all } e \in E. \qquad (6.1d)$$

Note that this formulation as a linear program only differs in equation (6.1c) from the linear programming formulation of the traditional maximum flow problem given in equations (2.1) on page 14. However, the known combinatorial algorithms for the traditional maximum flow problem cannot be applied directly to MFGPN. Instead, we need to make use of new approaches and generalizations of existing results. The following definitions build the basis for the theoretical framework that will be used in the remainder of this chapter.

Definition 6.6 (Types of edges):
Let x be a feasible flow in a generalized processing network. An edge $e = (v, w) \in E$ is said to be of *type* u if $x_e = u_e$. Similarly, if $x_e < u_e$ and $x_e = \alpha_e \cdot \sum_{e' \in \delta^+(v)} x_{e'}$, the edge is said to be of *type* α. ◁

Definition 6.7 ((Basic) Flow distribution scheme):
A function $\beta \colon E \to [0, 1]$ with $\beta_e := \beta(e) \leqslant \alpha_e$ for each $e \in E$ is called a *flow distribution scheme* if, for each $v \in V \setminus \{t\}$, it holds that $\sum_{e \in \delta^+(v)} \beta_e = 1$. Furthermore, if there is at most one edge $e \in \delta^+(v)$ with $\beta_e \notin \{0, \alpha_e\}$ at each node $v \in V \setminus \{t\}$, the function is called a *basic flow distribution scheme*. ◁

Intuitively, each flow distribution scheme determines how flow that arrives at some node $v \in V$ is sent through the outgoing edges without violating the dynamic capacity constraints or the flow conservation constraints. Thus, each flow distribution scheme together with a sufficiently small flow value val(x) determines a feasible flow x. Note that the concept of basic flow distribution schemes is a generalization of the notion of s-t-paths in traditional network flow problems since we obtain such a path for the case that $\alpha_e = 1$ for each $e \in E$. Moreover, note that the fraction $\frac{x_e}{\text{val}(x)}$ is constant for each edge $e \in E$ and every flow x determined by a given flow distribution scheme β, which leads to the following definition:

Definition 6.8 (Flow on flow distribution scheme, weight of edge in flow distribution scheme):
Let β be a flow distribution scheme. A flow x that fulfills $x_e = \beta_e \cdot \sum_{e' \in \delta^+(v)} x_{e'}$ for each $e = (v, w) \in E$ is called a *flow on* β. For a flow x on β with positive flow value val(x), the fraction $w_\beta(e) := \frac{x_e}{\text{val}(x)} \in [0, 1]$ (which is independent of the choice of x) is called the *weight of e in* β. ◁

In particular, note that the weight function w_β also embodies a flow with unit flow value for each flow distribution scheme β.

The notion of flow distribution schemes shows the main difference between our model and traditional processing networks: In the latter model, for each flow distribution scheme β, it always holds that $\beta_e = \alpha_e$ for each edge $e = (v, w)$ that leaves a special node v. This implies that the flow on each edge in $\delta^+(v)$ is determined by the flow on e. In our generalized model, however, there are multiple possible (basic) flow distribution schemes at each such node, which prevents us from directly applying the known algorithms for traditional processing networks to the generalized model.

6.3 Structural Results

We start by generalizing existing results for traditional flows to the case of MFGPN. As it turns out, a flow decomposition that is similar to the well-known flow decomposition of traditional flows is possible in the case of MFGPN as well (cf. (Ahuja et al., 1993)). To obtain this result, we need the following lemma:

Lemma 6.9:
A feasible flow on a given flow distribution scheme β that is positive on at least one edge can be determined in $\mathcal{O}(m^3)$ time.

Proof: Let $E_0 := E \cup \{e_0\}$ with $e_0 = (t, s)$ and $\beta_{e_0} := 1$. We show that we can find a non-zero feasible circulation[1] x on β (extended to e_0) in $G_0 = (V, E_0)$ within the given time bound, which clearly shows the claim.

Consider some edge $e = (v, w) \in E_0$. For every feasible circulation x on β, it must hold that $x_e - \beta_e \cdot \sum_{e' \in \delta^-(v)} x_{e'} = 0$ in order to be feasible on β and to fulfill flow conservation. The set of these constraints for each $e \in E$ builds a homogeneous linear equation system of the form $A \cdot x = 0$ over $m + 1$ variables with $m + 1$ constraints. Note that the sum of the coefficients amounts to zero in each row and column. Hence, the rank of the matrix A is at most m and, thus, the dimension of the kernel is at least one. Consequently, there is a non-zero vector x that solves the linear equation system. Without loss of generality, there is at least one edge $e = (v, w)$ with $x_e > 0$ in this vector. Since $x_e - \beta_e \cdot \sum_{e' \in \delta^-(v)} x_{e'} = 0$, it both holds that $\sum_{e' \in \delta^-(v)} x_{e'} > 0$ (i.e. there is at least one edge $e' \in \delta^-(v)$ with $x_{e'} > 0$) and that $x_{e''} \geqslant 0$ for each $e'' \in \delta^+(v)$ (since $x_{e''} - \beta_{e''} \cdot \sum_{e' \in \delta^-(v)} x_{e'} = 0$ as well and $\beta_{e''} \geqslant 0$). Since the underlying graph G_0 is strongly connected according to Assumption 6.4, Assumption 6.5, and due to the

---

1   A *feasible circulation* x is a feasible flow that fulfills $\text{excess}_x(v) = 0$ for each node $v \in V$.

additional edge e_0, an inductive argument yields that $x_e \geqslant 0$ for each $e \in E_0$. Hence, a suitable multiple of x yields a feasible flow in the underlying network. Since such a vector x can, e.g., be found by the Gaussian elimination procedure in $\mathcal{O}(m^3)$ time, the claim follows. $\qquad\square$

Theorem 6.10:
Each flow x can be decomposed into $\kappa \leqslant 2m$ flows $x^{(i)}$ on basic flow distribution schemes $\beta^{(i)}$ for $i \in \{1, \ldots, \kappa\}$. Such a decomposition can be found in $\mathcal{O}(m^4)$ time.

Proof: Let x be a feasible flow in $G = (V, E)$ with flow value $\mathrm{val}(x) > 0$. Without loss of generality, we can ignore edges carrying zero flow. For each $v \in V$ with positive outflow, let $(e_1, \ldots, e_k)$ denote an ordering of the edges in $\delta^+(v)$ such that $\frac{x_{e_i}}{\alpha_{e_i}} \geqslant \frac{x_{e_j}}{\alpha_{e_j}}$ for $i < j$ (in particular, edges of type α are located at the front of the ordering). If $\sum_{i=1}^{k} \alpha_{e_i} = 1$, we set $\beta_{e_i} := \alpha_{e_i}$ for each $i \in \{1, \ldots, k\}$. Else, if $\sum_{i=1}^{k} \alpha_{e_i} > 1$, there is some index $h \leqslant k$ with $\sum_{i=1}^{h-1} \alpha_{e_i} \leqslant 1$ and $\sum_{i=1}^{h} \alpha_{e_i} > 1$. By setting $\beta_{e_i} := \alpha_{e_i}$ for $i \in \{1, \ldots, h-1\}$, $\beta_{e_h} := 1 - \sum_{i=1}^{h} \beta_{e_i}$, and $\beta_{e_j} := 0$ for $j \in \{h+1, \ldots, k\}$ for each such node $v \in V$, we, thus, obtain a basic flow distribution scheme. Note that, for each $e \in E$, it holds that $\beta_e > 0$ only if $x_e > 0$ and that $\beta_e = \alpha_e$ whenever e is of type α.

Let $\overline{x}$ denote a feasible flow on β, which can be found in $\mathcal{O}(m^3)$ time according to Lemma 6.9. We claim that, for a suitable choice of $\delta > 0$, the flow $x(\delta) := x - \delta \cdot \overline{x}$ remains feasible. Obviously, for each choice of δ, flow conservation remains fulfilled at each node $v \in V$ since

$$\sum_{e \in \delta^-(v)} (x(\delta))_e - \sum_{e \in \delta^+(v)} (x(\delta))_e = \sum_{e \in \delta^-(v)} (x_e - \delta \cdot \overline{x}_e) - \sum_{e \in \delta^+(v)} (x_e - \delta \cdot \overline{x}_e)$$

$$= \left(\sum_{e \in \delta^-(v)} x_e - \sum_{e \in \delta^+(v)} x_e \right) - \delta \cdot \left(\sum_{e \in \delta^-(v)} \overline{x}_e - \sum_{e \in \delta^+(v)} \overline{x}_e \right) = 0 - 0 = 0.$$

For the flow on each edge $e \in E$ to remain non-negative, it must hold that $x((\delta))_e = x_e - \delta \cdot \overline{x}_e \geqslant 0$, i.e., $\delta \leqslant \frac{x_e}{\overline{x}_e}$ for each $e \in E$ with $\overline{x}_e > 0$. Moreover, in order to fulfill the dynamic capacity of each edge $e = (v, w) \in E$, the value δ must fulfill the following constraint:

$$(x(\delta))_e \leqslant \alpha_e \cdot \sum_{e' \in \delta^+(v)} (x(\delta))_{e'}$$

$$\Longleftrightarrow \quad (x_e - \delta \cdot \overline{x}_e) \leqslant \alpha_e \cdot \sum_{e' \in \delta^+(v)} (x_{e'} - \delta \cdot \overline{x}_{e'})$$

$$\Longleftrightarrow \quad \delta \cdot \left(\alpha_e \cdot \left(\sum_{e' \in \delta^+(v)} \overline{x}_{e'} \right) - \overline{x}_e \right) \leqslant \alpha_e \cdot \left(\sum_{e' \in \delta^+(v)} x_{e'} \right) - x_e. \qquad (6.2)$$

Note that both sides of inequality (6.2) are non-negative since $\bar{x}$ and x fulfill the dynamic capacity constraints. For the case that e is of type α in $\bar{x}$ (which is, e.g., true if e is also of type α in x according to the construction of β), inequality (6.2) is fulfilled for every choice of δ since $\alpha_e \cdot \left(\sum_{e' \in \delta^+(v)} \bar{x}_{e'} \right) - \bar{x}_e = 0$. Otherwise, (6.2) is equivalent to

$$\delta \leqslant \frac{\alpha_e \cdot \left(\sum_{e' \in \delta^+(v)} x_{e'} \right) - x_e}{\alpha_e \cdot \left(\sum_{e' \in \delta^+(v)} \bar{x}_{e'} \right) - \bar{x}_e}. \tag{6.3}$$

Let δ be the maximum value that fulfills both $\delta \leqslant \frac{x_e}{\bar{x}_e}$ for each e with $\bar{x}_e > 0$ and inequality (6.3) for each $e \in E$ that is not of type α in $\bar{x}$. By the above arguments, it follows that $\delta \cdot \bar{x}$ is a feasible flow on β and that the remaining flow $x(\delta)$ is feasible as well. Moreover, note that the flow on at least one edge in $x(\delta)$ becomes zero (for the case that $\delta = \frac{x_e}{\bar{x}_e}$ for some edge $e \in E$), or at least one edge $e \in E$ that was not of type α in x is of type α in $x(\delta)$. In the latter case, edge e will remain of type α after each of the following iterations of the algorithm according to the definition of β.

Hence, the above procedure executes at most $2m$ iterations while each of these iterations runs in $\mathcal{O}(m^3)$ time according to Lemma 6.9, which shows the claim. $\square$

Note that, on a graph without dynamic capacities (i.e., with $\alpha_e = 1$ for each $e \in E$), each flow on a basic flow distribution scheme β either corresponds to a flow on an s-t-path or on a cycle. Thus, Theorem 6.10 is a generalization of the flow decomposition theorem for traditional flows (cf. (Ahuja et al., 1993)).

We now restrict our considerations to the case of acyclic graphs. In this case, the running time of finding a flow decomposition can be significantly improved compared to Theorem 6.10. Recall that the weight $w_\beta(e)$ of an edge $e \in E$ in a flow distribution scheme β is independent of the choice of the underlying flow x according to Definition 6.8.

Lemma 6.11:
The weights $w_\beta(e)$ of all edges $e \in E$ in a given flow distribution scheme β can be determined in $\mathcal{O}(m)$ time on acyclic graphs.

Proof: Let $(v_1 = s, v_2, \ldots, v_{n-1}, v_n = t)$ denote a topological sorting of the nodes, which can be determined in $\mathcal{O}(m)$ time (cf., e.g., Cormen et al. (2009)). For $i = 1$, the weight of each edge $e \in \delta^+(v_i)$ is directly given by β, i.e., we get that $w_\beta(e) := \beta_e$. Now assume that we know the weights for all edges in $\delta^+(v_j)$ for $j \in \{1, \ldots, i\}$ and consider the subsequent node v_{i+1} in the ordering. In each flow x on β with flow value $val(x)$, the amount of flow that reaches v_{i+1} is given by $F := \sum_{e \in \delta^-(v_{i+1})} x_e =$

$\sum_{e \in \delta^-(v_{i+1})} \text{val}(x) \cdot w_\beta(e)$. For each $e \in \delta^+(v_{i+1})$, the flow on x_e is then given by $x_e = \beta_e \cdot F$, i.e., the weight of e amounts to

$$w_\beta(e) = \frac{x_e}{\text{val}(x)} = \frac{\beta_e \cdot F}{\text{val}(x)} = \beta_e \cdot \sum_{e' \in \delta^-(v_{i+1})} w_\beta(e').$$

Repeating the above procedure for each node $v_i \in V$, the weight of each edge in β can be determined in $\mathcal{O}(m)$ time, which shows the claim. $\square$

Corollary 6.12:
A feasible flow on a given flow distribution scheme β that is positive on at least one edge can be determined in $\mathcal{O}(m)$ time on acyclic graphs.

Proof: According to Lemma 6.11, we can determine the weights $w_\beta(e)$ in β of all edges $e \in E$ in $\mathcal{O}(m)$ time. Note that a flow of value F on β results in a flow of value $w_\beta(e) \cdot F$ on each edge $e \in E$. Hence, for any F with $0 < F \leqslant \min\{\frac{u_e}{w_\beta(e)} : e \in E$ and $w_\beta(e) > 0\}$, the flow x with $x_e := F \cdot w_\beta(e)$ for each $e \in E$ is a feasible flow with positive flow value F, which shows the claim. $\square$

Using the result of Corollary 6.12 in the proof of Theorem 6.10, we immediately get the following result:

Theorem 6.13:
On acyclic graphs, each flow x can be decomposed into at most $2m$ flows on basic flow distribution schemes in $\mathcal{O}(m^2)$ time. $\square$

6.4 Complexity and Approximability

In this section, we consider the complexity and approximability of the maximum flow problem in generalized processing networks. Although MFGPN is solvable in polynomial time, it turns out to be much harder to solve than the maximum flow problem in traditional networks. Nevertheless, for the case of acyclic graphs, we will be able to derive an FPTAS for the problem that runs in strongly polynomial time.

6.4.1 Complexity

Note that the linear program (6.1a) – (6.1d) can be solved in (weakly) polynomial time by known techniques such as interior point methods (cf. (Schrijver, 1998)). In particular, using the procedure by Vaidya (1989) that was described in Section 4.2, we get the following weakly polynomial running time for MFGPN:

Theorem 6.14:

MFGPN is solvable in weakly polynomial time $\mathcal{O}\left(m^{3.5} \log M\right)$ if $M \geqslant \max_{e \in E} u_e$ and each flow ratio is a rational number with numerator and denominator at most M. $\quad\square$

However, as in the previous chapters of this thesis, we are in particular interested in combinatorial algorithms for the treated problems that exploit the discrete structure of the underlying network. As for the case of traditional flows, the flow decomposition theorem that was derived in Section 6.3 is only a structural result and does not immediately yield an algorithm that solves the problem of finding an optimal solution. In fact, it turns out that the problem MFGPN seems to be much more complicated than the traditional maximum flow problem since every packing LP of the form $\max\left\{c^T x : A x \leqslant b, x \geqslant 0\right\}$ for positive rational vectors c and b and a matrix A with non-negative entries can be reduced to MFGPN in linear time. This result was first published by Schaefer (1978). Since it seems to be widely unnoticed in present literature, we present a short proof in the following.

Theorem 6.15 (Schaefer (1978)):

Every packing LP can be solved by computing a maximum flow in a generalized processing network. This network can be constructed from the given packing LP in linear time.

Proof: Let $\max\left\{c^T x : A^T x \leqslant b, x \geqslant 0\right\}$ be a packing LP with $c \in \mathbb{Q}^n$, $b \in \mathbb{Q}^m$, $A \in \mathbb{Q}^{m \times n}$, $c_j > 0$ for $j \in \{1, \ldots, n\}$, $b_i > 0$ for $i \in \{1, \ldots, m\}$, and $a_{ij} \geqslant 0$ for $j \in \{1, \ldots, n\}, i \in \{1, \ldots, m\}$.

Without loss of generality, we may assume that $b_i = 1$ for each $i \in \{1, \ldots, m\}$ and that $c_j \geqslant 1$ for each $j \in \{1, \ldots, n\}$ since we can otherwise scale the corresponding row or the objective function, respectively, by appropriate factors. Similarly, we may assume that $\sum_{i=1}^m a_{ij} \leqslant 1$ for each $j \in \{1, \ldots, n\}$: Otherwise, for $1 < q :=$ $\max\left\{\sum_{i=1}^m a_{ij} : j \in \{1, \ldots, n\}\right\}$, we could use the equivalent LP formulation $\max\{\frac{1}{q} c^T x' : \frac{1}{q} A^T x' \leqslant 1, x' \geqslant 0\}$ and afterwards substitute $x := \frac{1}{q} \cdot x'$.

We construct an instance of MFGPN as follows: Aside from a source s and sink t, we insert two nodes v_j and v_j' for each $j \in \{1, \ldots, n\}$ and a node w_i for each $i \in \{1, \ldots, m\}$. We connect s with each node v_j and insert an edge with capacity 1 between each node w_i and the sink t. Moreover, we insert an edge between v_j and v_j' with flow ratio $\frac{1}{c_j}$ and an edge that heads from v_j to t with flow ratio $1 - \frac{1}{c_j}$. Finally, we add an edge between each v_j' and each w_i with flow ratio a_{ij} and one edge between each v_j' and t with flow ratio $1 - \sum_{i=1}^m a_{ij}$. If not mentioned explicitly, the flow ratio of each edge is 1 and the capacity is infinite. An example of a packing LP and the corresponding network is depicted in Figure 6.1.

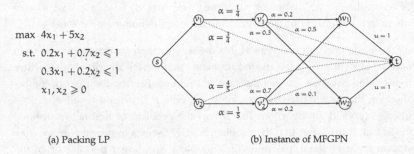

$$\max \quad 4x_1 + 5x_2$$
$$\text{s.t.} \quad 0.2x_1 + 0.7x_2 \leqslant 1$$
$$0.3x_1 + 0.2x_2 \leqslant 1$$
$$x_1, x_2 \geqslant 0$$

(a) Packing LP   (b) Instance of MFGPN

Figure 6.1: An example packing LP (left) and the corresponding instance of MFGPN (right). If not depicted, the flow ratio of each edge is 1 and the capacity is infinite.

It is now easy to see that the flow value of a maximum flow in the constructed network equals the optimum value of the given (transformed) packing LP instance: For $j \in \{1, \ldots, n\}$, the flow on the edge between v_j and v_j' can be interpreted as the value of variable x_j. Since this edge has a flow ratio of $\frac{1}{c_j}$, it is necessary to send $c_j \cdot x_j$ units of flow from s to v_j in order to obtain a flow of value x_j between v_j and v_j', which corresponds to the contribution of x_j to the objective function value of the LP. Moreover, for $i \in \{1, \ldots, m\}$, the flow on the edge between w_i and t can be interpreted as the left-hand side of the constraint $\sum_{j=1}^{n} a_{ij} \cdot x_j \leqslant 1$ and the capacity of the edge enforces the constraint to be fulfilled. Finally, the interconnections between the nodes v_j' and w_i model the effect that the corresponding variables x_j have on the value of the left-hand side of each constraint i.

Hence, by solving the constructed instance of MFGPN and interpreting the flow values on each edge between v_j and v_j' as the value of variable x_j, we obtain an optimal solution of the given packing LP. Since the above transformations and construction of the network work in linear time, the claim follows. □

6.4.2 Approximability

The results of the previous subsection imply that, even on acyclic graphs, the maximum flow problem in generalized processing networks is much more complicated than the maximum flow problem in traditional networks, so strongly polynomial-time combinatorial algorithms may not necessarily exist for MFGPN. Nevertheless, as it will be shown in the following, we can use the special structure of acyclic graphs in order to obtain an FPTAS for finding a maximum flow in an acyclic generalized processing network by incorporating the generalized packing framework that was introduced in Section 3.3. Even more, this FPTAS can be implemented to run in strongly polynomial time, in contrast to interior point methods. To this end, we need the following result:

Lemma 6.16:
Let y denote a function that assigns a positive weight $y_e := y(e) > 0$ to each edge $e \in E$. A basic flow distribution scheme β that minimizes the total weight $\sum_{e \in E} w_\beta(e) \cdot y_e$ can be found in $\mathcal{O}(m)$ time on acyclic graphs.

Proof: Let $(v_1, \ldots, v_n)$ denote a topological sorting of the node set V, which can be found in $\mathcal{O}(m)$ time. For each $i \in \{1, \ldots, n\}$, let $G^{(i)} := (V^{(i)}, E^{(i)})$ with $V^{(i)} := \{v_i, \ldots, v_n\}$ and $E^{(i)} := \{e = (v_j, v_l) \in E : j, l \geqslant i\}$ denote the subgraph induced by $\{v_i, \ldots, v_n\}$. Moreover, let $w(i)$ denote the minimum total weight of a basic flow distribution scheme in $G^{(i)}$ with respect to y.

Clearly, since $G^{(n)}$ contains no edge at all, it holds that $w(n) = 0$. Now assume that we want to determine the value of $w(i)$ for some $i \in \{1, \ldots, n-1\}$ and that the values $w(i+1), \ldots, w(n)$ are already known. In order to find a (not necessarily basic) flow distribution scheme β, we need to assign values $\beta_e \in [0, 1]$ to each $e \in \delta^+(v_i)$ such that $\sum_{e \in \delta^+(v_i)} \beta_e = 1$. A value of β_e for some edge $e = (v_i, v_l) \in \delta^+(v_i)$ increases the total weight $w(i)$ by $\beta_e \cdot y_e + \beta_e \cdot w(l)$ since $w_\beta(e) = \beta_e$ in $G^{(i)}$ and since a fraction β_e of the total flow must be sent through $G^{(l)}$. Thus, the minimum total weight in $G^{(i)}$ is given by the following linear program:

$$w(i) = \min \sum_{e = (v_i, v_l) \in \delta^+(v_i)} \beta_e \cdot (y_e + w(l))$$

$$\text{s.t.} \sum_{e \in \delta^+(v_i)} \beta_e = 1,$$

$$0 \leqslant \beta_e \leqslant \alpha_e \qquad \text{for all } e \in E.$$

Similarly to the fractional knapsack problem (cf. Kellerer et al. (2004)), it is easy to see that an optimal solution to this fractional packing problem can be determined by the following procedure: If $\sum_{e \in \delta^+(v_i)} \alpha_e = 1$, the only feasible solution is given by $\beta_e = \alpha_e$ for each $e \in \delta^+(v_i)$. Otherwise, if $\sum_{e \in \delta^+(v_i)} \alpha_e > 1$, let $(e_1, \ldots, e_k)$ denote a sorting of the outgoing edges of v_i in non-decreasing order of their coefficients $y_e + w(l)$. Let l be the unique index such that $\sum_{j=1}^{l-1} \alpha_e \leqslant 1$ and $\sum_{j=1}^{l} \alpha_e > 1$. By setting $\beta_{e_j} := \alpha_{e_j}$ for $j \in \{1, \ldots, l-1\}$, $\beta_{e_l} = 1 - \sum_{j=1}^{l-1} \beta_{e_j}$, and $\beta_{e_j} = 0$ for $j \in \{l+1, \ldots, k\}$, we then get an optimal solution. Similar to the fractional knapsack problem, we can find this index l

in $\mathcal{O}(k)$ time by using weighted medians (cf. (Korte and Vygen, 2002)). Note that, in this solution, it holds that $\beta_e \notin \{0, \alpha_e\}$ for at most one edge, i.e., the optimal solution is a basic flow distribution scheme. □

Note that the above algorithm is strongly combinatorial according to the definition that was given in Section 3.1. This leads to the following theorem:

Theorem 6.17:
There is an FPTAS for the maximum flow problem in acyclic generalized processing networks that runs in $\mathcal{O}\left(\frac{1}{\varepsilon^2} \cdot m^2 \log m\right)$ time.

Proof: The proof is composed of three results that have already been shown before: According to Theorem 6.10 (and Theorem 6.13), each flow x in a generalized processing network can be decomposed into at most $2m$ flows on basic flow distribution schemes, i.e., each flow x lies in the cone C that is generated by the (possibly exponential-size, but finite) set $S := \{w_\beta : \beta \text{ is a flow distribution scheme}\}$ of flows with unit flow value on basic flow distribution schemes. Moreover, since we can rewrite the objective function of the maximum flow problem in generalized processing networks as $\max \sum_{e \in E} c_e \cdot x_e$ with $c_e := 1$ for $e \in \delta^-(t)$ and $c_e := 0$ for $e \in E \setminus \delta^-(t)$, it holds that $\sum_{e \in E} c_e \cdot w_\beta(e) = 1$ for all flows $w_\beta \in S$. Hence, we obtain the following equivalent formulation of MFGPN:

$$\max \sum_{e \in E} c_e \cdot x_e$$
$$\text{s.t. } x_e \leqslant u_e \qquad\qquad \text{for all } e \in E,$$
$$x \in C.$$

The constraint matrix of this formulation only contains $N = m$ non-zero entries since both the flow conservation constraints and the dynamic capacity constraints are modeled by the containment in the cone C. Since, for a given cost vector y, we can determine a flow distribution scheme β that minimizes the total weight $\sum_{e \in E} w_\beta(e) \cdot y_e$ in $\mathcal{O}(m)$ time according to Lemma 6.16, the claim immediately follows by Theorem 3.5. □

Note that Theorem 6.17 can be easily generalized to the *minimum-cost flow problem in a generalized processing network*, in which the objective function is replaced by a general linear cost function of the form $\min \sum_{e \in E} c_e \cdot x_e$. By the same arguments that were used in the proof of Theorem 6.17, we get an FPTAS for this much more general problem running in $\mathcal{O}\left(\frac{1}{\varepsilon^2} \cdot m^3 \log m\right)$ time by using Theorem 3.10. Moreover, note that we can solve budget-constrained versions of both the maximum and the minimum cost flow problem in a generalized processing network within the same running times as the unconstrained versions according to Theorem 3.5 and Theorem 3.10, respectively.

6.5 Series–Parallel Graphs

In this section, we investigate the maximum flow problem for generalized processing networks on series-parallel graphs. Since each series-parallel graph is acyclic, in particular, the positive results from Section 6.4 apply here as well. However, we are now able to derive two algorithms that compute a maximum flow in a series-parallel generalized processing network in (strongly) polynomial time. In the following three results, we investigate a special case of the problem which will be used as a building block for the two upcoming polynomial-time procedures in Section 6.5.1 and Section 6.5.2. To this end, let $E(v, w) := \delta^+(v) \cap \delta^-(w)$ denote the set of all edges between the two nodes $v, w \in V$.

Lemma 6.18:
Let v and w be two nodes such that all edges $\{e_1, \ldots, e_k\}$ that leave v are parallel edges heading to w, i.e., $\delta^+(v) = E(v, w)$, and assume that the edges are ordered such that $\frac{u_{e_i}}{\alpha_{e_i}} \leqslant \frac{u_{e_j}}{\alpha_{e_j}}$ for $i < j$. Then the maximum flow between v and w fulfills the property that there exists an index $h \in \{1, \ldots, k\}$ such that all edges e_i with $1 \leqslant i \leqslant h$ are of type u and all edges e_j with $h + 1 \leqslant j \leqslant k$ are of type α. This index h can be computed in $\mathcal{O}(k)$ time.

Proof: Let x be any maximum flow between v and w. Clearly, we may assume that each edge is either of type u or of type α in x since the total flow value could else be further improved. In the following, we show that we can label several edges with type α in an iterative process from right to left (i.e., from higher indices to lower indices) until we find the desired index h, which allows us to assign the remaining edges to type u and stop the procedure.

Let e_i be some edge that has not yet been labeled such that all edges e_j with $i + 1 \leqslant j \leqslant k$ are of type α (initially, choose $i := k$). Since these edges e_j are of type α, a fixed fraction $\alpha^{(i)} := \sum_{j=i+1}^{k} \alpha_{e_j}$ of the total outflow of v flows through the edges e_j, $i + 1 \leqslant j \leqslant k$. Thus, the maximum flow value $F := \mathrm{val}(x) = \sum_{j=1}^{k} x_{e_j}$ is determined by the flow values on the edges $e_1, \ldots, e_i$ as

$$F = \sum_{j=1}^{k} x_{e_j} = \sum_{j=1}^{i} x_{e_j} + \sum_{j=i+1}^{k} x_{e_j} = \sum_{j=1}^{i} x_{e_j} + \alpha^{(i)} \cdot F$$

$$\iff F = \frac{1}{1 - \alpha^{(i)}} \cdot \sum_{j=1}^{i} x_{e_j}.$$

First consider the case that $\alpha_{e_i} < (1 - \alpha^{(i)}) \cdot \frac{u_{e_i}}{\sum_{j=1}^i u_{e_j}}$. In this case, edge e_i cannot be of type u since this would imply that

$$\alpha_{e_i} \cdot F = \alpha_{e_i} \cdot \frac{1}{1 - \alpha^{(i)}} \cdot \sum_{j=1}^i x_{e_j} \leqslant \alpha_{e_i} \cdot \frac{1}{1 - \alpha^{(i)}} \cdot \sum_{j=1}^i u_{e_j} < u_{e_i} = x_{e_i},$$

so the dynamic capacity of edge e_i would be violated. Thus, edge e_i and, hence, all edges e_j with $j \in \{i, \ldots, k\}$ are of type α.

Now consider the case that $\alpha_{e_i} \geqslant (1 - \alpha^{(i)}) \cdot \frac{u_{e_i}}{\sum_{j=1}^i u_{e_j}}$. By setting $x'_{e_j} := u_{e_j}$ for $j \in \{1, \ldots, i\}$ and $x'_{e_l} := x_{e_l}$ for $l \in \{i+1, \ldots, k\}$, the dynamic capacity constraint of each edge e_j is fulfilled for x' since

$$\alpha_{e_j} \cdot F = \alpha_{e_j} \cdot \frac{1}{1 - \alpha^{(i)}} \cdot \sum_{l=1}^i x_{e_l} = \frac{\alpha_{e_j}}{u_{e_j}} \cdot u_{e_j} \cdot \frac{1}{1 - \alpha^{(i)}} \cdot \sum_{l=1}^i u_{e_l}$$

$$\geqslant \frac{\alpha_{e_i}}{u_{e_i}} \cdot u_{e_j} \cdot \frac{1}{1 - \alpha^{(i)}} \cdot \sum_{l=1}^i u_{e_l} \geqslant u_{e_j} = x'_{e_j}.$$

Thus, by maximality of x, we must have $x_{e_j} = u_{e_j}$ for $j \in \{1, \ldots, i\}$. In total, by setting $h := i$, we can label each edge e_j with $j \in \{1, \ldots, h\}$ with type u and each edge e_l for $l \in \{h+1, \ldots, k\}$ with type α, which shows the claim. $\square$

Corollary 6.19:
Let v and w be two nodes such that all edges $\{e_1, \ldots, e_k\}$ that leave v are parallel edges heading to w, i.e., $\delta^+(v) = E(v, w)$, and assume that the edges are ordered such that $\frac{u_{e_i}}{\alpha_{e_i}} \leqslant \frac{u_{e_j}}{\alpha_{e_j}}$ for $i < j$. The maximum flow between v and w can be found in $\mathcal{O}(k)$ time.

Proof: According to Lemma 6.18, there exists a maximum flow x and an index h such that each edge e_i with $1 \leqslant i \leqslant h$ is of type u and each edge e_j with $h+1 \leqslant j \leqslant k$ is of type α in x and this index h can be computed in $\mathcal{O}(k)$ time. The flow x_{e_i} on the edges e_i is consequently given by $x_{e_i} := u_{e_i}$ for $1 \leqslant i \leqslant h$. Since a fixed fraction $\alpha^{(h)} = \sum_{j=h+1}^k \alpha_{e_j}$ of the total flow is sent along the edges e_j for $h+1 \leqslant j \leqslant k$, the total flow value is given by $F := \frac{1}{1-\alpha^{(h)}} \cdot \sum_{i=1}^h u_{e_i}$ and the flow on each edge e_j amounts to $x_{e_j} := \alpha_{e_j} \cdot F$. $\square$

As seen in Lemma 6.18, there is some index h such that all edges e_i with $1 \leqslant i \leqslant h$ are of type u and the remaining edges are of type α in a maximum flow. In the following lemma, we show that the converse is true as well. In particular, this shows that the maximum flow is unique. Note that this lemma considers the more general case, in which the edges in $\delta^+(v)$ are not assumed to be necessarily parallel:

Lemma 6.20:
Let v be a node such that, for some given feasible flow x, at least one of the outgoing edges $\delta^+(v)$ of v is of type u while the rest of the edges is of type α. Then the flow on the edges in $\delta^+(v)$ is unique and maximum.

Proof: As in Lemma 6.18, assume that the edges $\{e_1, \ldots, e_k\}$ are ordered such that $\frac{u_{e_i}}{\alpha_{e_i}} \leqslant \frac{u_{e_j}}{\alpha_{e_j}}$ for $i < j$. Since all edges in $\delta^+(v)$ are either of type u or of type α in the flow x, the same arguments that were used in the proof of Lemma 6.18 show that there is some index $h \in \{1, \ldots, k\}$ with $\alpha_{e_h} \geqslant (1 - \alpha^{(h)}) \cdot \frac{u_{e_h}}{\sum_{l=1}^h u_{e_l}}$ such that none of the edges e_j for $j \in \{h+1, \ldots, k\}$ can be of type u and is, thus, of type α. Furthermore, since $\frac{u_{e_i}}{\alpha_{e_i}} \leqslant \frac{u_{e_h}}{\alpha_{e_h}}$ for each $i \in \{1, \ldots, h\}$, we get that

$$\alpha_{e_i} \geqslant \frac{u_{e_i}}{u_{e_h}} \cdot \alpha_{e_h} \geqslant \frac{u_{e_i}}{u_{e_h}} \cdot (1 - \alpha^{(h)}) \cdot \frac{u_{e_h}}{\sum_{l=1}^h u_{e_l}} = (1 - \alpha^{(h)}) \cdot \frac{u_{e_i}}{\sum_{l=1}^h u_{e_l}}.$$

Let $I_\alpha := \{i \in \{1, \ldots, h\} : e_i \text{ is of type } \alpha\}$ and $I_u := \{i \in \{1, \ldots, h\} : e_i \text{ is of type } u\}$, where $I_u \neq \emptyset$ by assumption. The flow value F out of node v is then given by

$$F = \sum_{i=1}^k x_{e_i} = \sum_{j=h+1}^k \alpha_{e_j} \cdot F + \sum_{i \in I_\alpha} \alpha_{e_i} \cdot F + \sum_{i \in I_u} u_{e_i},$$

which is equivalent to

$$F = \frac{1}{1 - \alpha^{(h)} - \sum_{i \in I_\alpha} \alpha_{e_i}} \cdot \sum_{i \in I_u} u_{e_i}$$

$$\geqslant \frac{1}{1 - \alpha^{(h)} - \sum_{i \in I_\alpha} (1 - \alpha^{(h)}) \cdot \frac{u_{e_i}}{\sum_{l=1}^h u_{e_l}}} \cdot \sum_{i \in I_u} u_{e_i}$$

$$= \frac{1}{(1 - \alpha^{(h)}) \cdot \left(1 - \frac{\sum_{i \in I_\alpha} u_{e_i}}{\sum_{l=1}^h u_{e_l}}\right)} \cdot \sum_{i \in I_u} u_{e_i}$$

$$= \frac{1}{(1 - \alpha^{(h)}) \cdot \frac{\sum_{i \in I_u} u_{e_i}}{\sum_{l=1}^h u_{e_l}}} \cdot \sum_{i \in I_u} u_{e_i} = \frac{1}{1 - \alpha^{(h)}} \cdot \sum_{l=1}^h u_{e_l}.$$

Note that this lower bound on the flow value equals the flow value that is given by setting $x_{e_i} := u_{e_i}$ for each $i \in \{1, \ldots, h\}$. Thus, since the given flow is feasible, each of the edges e_i for $i \in \{1, \ldots, h\}$ must be of type u, which is clearly maximum and shows the claim. $\qquad \square$

6.5.1 Augmenting on Flow Distribution Schemes

In this subsection, we describe an algorithm that repeatedly sends flow on flow distribution schemes with positive residual capacity, i.e., on which a given flow can be increased without violating any capacity constraint. This algorithm is similar to the well-known augmenting path algorithm for the traditional maximum flow problem, but generalizes the procedure from augmentations on single paths to augmentations on (basic) flow distribution schemes. As it turns out, it is possible to augment flow in a greedy manner without the need to use some form of residual network in order to revert prior decisions, which allows us to obtain a strongly polynomial-time algorithm for the problem on series-parallel graphs that runs in $\mathcal{O}(m^2)$ time. Note that a similar result is known for the traditional maximum flow and minimum cost flow problem in series-parallel graphs (Bein et al., 1985).

The result will be established in four steps: We first show that we can find a suitable starting solution efficiently (Lemma 6.21). We then define a measurement α_x that allows us to evaluate easily if or if not there is an augmenting flow distribution scheme (Lemma 6.22). In a next step, we prove that the procedure terminates within $2m$ augmentations (Corollary 6.28). Finally, we show that the resulting flow is in fact maximal (Theorem 6.31).

As a starting flow for our algorithm, we use a flow that is positive on each edge. Such a flow can be found in linear time even on general acyclic graphs, as the following lemma shows:

Lemma 6.21:
Let G be an acyclic graph. In $\mathcal{O}(m)$ time, we can compute a feasible flow x that is positive on each edge and that fulfills the property that, for each $v \in V \setminus \{t\}$, either all or no edges in $\delta^+(v)$ are of type α and no edge is of type u.

Proof: For each $e = (v, w) \in E$, let $\beta_e := \frac{\alpha_e}{\sum_{e' \in \delta^+(v)} \alpha_{e'}}$. Clearly, β is a feasible flow distribution scheme that assigns a positive value β_e to each edge. Moreover, for each $v \in V \setminus \{t\}$, if $\sum_{e' \in \delta^+(v)} \alpha_{e'} = 1$, then $\beta_e = \alpha_e$ for all $e \in \delta^+(v)$ (so all edges in $\delta^+(v)$ are of type α in any flow on β). Otherwise, $\sum_{e' \in \delta^+(v)} \alpha_{e'} > 1$ and we obtain that $\beta_e < \alpha_e$ for each $e \in \delta^+(v)$ (so no edge in $\delta^+(v)$ is of type α in any flow on β). According to Lemma 6.11, we can compute the corresponding (positive) weights $w_\beta(e)$ of all edges in $\mathcal{O}(m)$ time. Using these weights, we obtain a feasible flow x with the desired properties by setting $x_e := w_\beta(e) \cdot F$ for $0 < F < \min\left\{\frac{u_e}{w_\beta(e)} : e \in E\right\}$. $\square$

After sending a small amount of flow that is positive on each edge as described in Lemma 6.21, we may assume in the following that the values α_e of all outgoing edges of type α sum up to at most one at each node, which may not be true for the zero-flow[2].

In the following, for a given series-parallel graph G and a feasible flow x in G, we call a flow distribution scheme β *augmenting* if we can increase x by adding some flow x' on β of positive flow value without violating any static or dynamic capacity constraint. Clearly, if we can find an augmenting flow distribution scheme β, the flow x cannot be optimal since we are able to increase the flow value by sending flow on β. As we will see in the following, the reverse is true as well, i.e., as soon as there is no further augmenting flow distribution scheme, the flow is maximum.

Given a flow x, we start by defining a function $\alpha_x(G')$ that measures the maximum fraction of augmenting flow that can be sent through each subgraph G' corresponding to a node in the decomposition tree of the given series-parallel network G:

$$\alpha_x(G' = e) = \begin{cases} 0, & \text{if } e \text{ is of type u in } x, \\ \alpha_e, & \text{if } e \text{ is of type } \alpha \text{ in } x, \\ 1, & \text{else.} \end{cases} \tag{6.4a}$$

$$\alpha_x(G' = G_1 \mid G_2) = \min\{1, \alpha_x(G_1) + \alpha_x(G_2)\}. \tag{6.4b}$$

$$\alpha_x(G' = G_1 \circ G_2) = \begin{cases} \alpha_x(G_1), & \text{if } \alpha_x(G_2) = 1, \\ 0, & \text{else.} \end{cases} \tag{6.4c}$$

Clearly, all values $\alpha_x(G')$ can be computed in $\mathcal{O}(m)$ time using a dynamic programming scheme on the decomposition tree of G. The following lemma and the resulting corollary show that we can use $\alpha_x(G)$ in order to decide whether an augmenting flow distribution scheme exists:

Lemma 6.22:
For a given flow x in a series-parallel graph G, it holds that $\alpha_x(G') = q$ for a given subgraph G' corresponding to a node in the composition tree of G and for some $q \in [0, 1]$ if and only if q is the maximum value in $[0, 1]$ such that a fraction q of a sufficiently small amount of additional flow that arrives at the source node of G' can be sent through G'.

Proof: We show the claim in a bottom-up manner by induction on the decomposition tree of G. Consider a leaf of the tree that corresponds to an edge $e = (v, w)$. Obviously, if e is of type u, the flow on e cannot be increased. If the edge is of type α, at most a

---

2   Clearly, this does not imply that the values α_e of *all* outgoing edges of some node v sum up to at most one, in contrast to the case of traditional processing networks.

fraction α_e of additional flow that arrives at v can be sent through e without violating the dynamic capacity constraint of the edge. Otherwise, if neither of the two capacities of e is reached, we can send all of the additional flow that arrives at v to w using edge e until e becomes of type u or type α. This behavior is modeled by equation (6.4a).

Now assume that G' with source s' is the parallel composition of the two series-parallel components G_1 and G_2, which can carry at most a fraction $\alpha_x(G_1) =: q_1$ and $\alpha_x(G_2) =: q_2$ of additional flow at s' by the induction hypothesis, respectively. Since flow that it sent through G' splits into two fractions that pass G_1 and G_2, respectively, the fraction of additional flow that can flow through G' amounts to at most $\min\{1, q_1 + q_2\}$, as given by equation (6.4b).

Finally, let G' with source s' be the series composition of two components G_1 and G_2 with common node v. Assume that G' can carry a positive fraction $q > 0$ of additional flow that arrives at s'. This is possible if and only if G_1 can carry a fraction of q (i.e., if $\alpha_x(G_1) \geqslant q$) and, since all the edges in $\delta^+(v)$ are already contained in G_2, the component G_2 can carry all the flow that arrives at v (i.e., if $\alpha_x(G_2) = 1$). Thus, according to equation (6.4c), $\alpha_x(G') = q$ in this case. Similarly, it holds that $\alpha_x(G') = 0$ if and only if $\alpha_x(G_1) = 0$ or $\alpha_x(G_2) < 1$, which is true if and only if G_1 cannot carry flow at all or G_2 cannot carry all of the flow that arrives at v, i.e., we cannot send a fraction of additional flow that arrives at s' through G'. □

Corollary 6.23:
There exists an augmenting flow distribution scheme in a series-parallel graph G with a given flow x if and only if $\alpha_x(G) = 1$. □

In the following, for a fixed decomposition tree of the series-parallel graph G, we define a total order $\prec$ on the edges with $e \prec e'$ if and only if there is an inner node in the decomposition tree such that e is reachable via the left and e' is reachable via the right child of the inner node. For example, in Figure 2.1 on page 11, it holds that $e_1 \prec e_3$ and that $e_4 \prec e_6$, but not that $e_4 \prec e_2$.

Lemma 6.24:
For a given flow x in a series-parallel graph G satisfying $\alpha_x(G) = 1$, it is possible to find an augmenting (basic) flow distribution scheme in $\mathcal{O}(m)$ time.

Proof: We show that we can recursively compute a *partial* augmenting basic flow distribution scheme $\beta(G', q)$ for each series-parallel graph G' corresponding to a node in the decomposition tree of G with source s' in $\mathcal{O}(m)$ time. This flow distribution scheme describes how a fraction q with $0 \leqslant q \leqslant \alpha_x(G')$ of the flow that leaves s' is routed through G'. Evaluating $\beta(G, 1)$ then gives the desired flow distribution scheme in $\mathcal{O}(m)$ time.

For a component consisting of a single edge $e = (s', t')$, the only possible partial flow distribution scheme that fulfills the required properties is given by $\beta_e := q$. Now let G' be the series composition of two series-parallel graphs G_1 and G_2. Similar to the definition of α_x in equation (6.4c), we need to evaluate both $\beta(G_1, q)$ and $\beta(G_2, 1)$ in order to distribute a fraction of q of the flow arriving at s' through G_1 and G_2. Finally, let G' be the parallel composition of the two series-parallel graphs G_1 and G_2 where G_1 is the left child of the inner node that corresponds to G' in the decomposition tree of G. We distribute the fraction q to both components by evaluating $\beta(G_1, q_1)$ and $\beta(G_2, q_2)$ with $q_1 := \min\{q, \alpha_x(G_1)\}$ and $q_2 := q - q_1$. It is now easy to see that starting the recursive procedure from the root of the decomposition tree by evaluating $\beta(G, 1)$ yields the desired augmenting flow distribution scheme in $\mathcal{O}(m)$ time. $\qquad\square$

In summary, for a given flow x in a series-parallel graph G, we are able to determine in $\mathcal{O}(m)$ time if the flow x can be improved by sending additional flow on some flow distribution scheme. If so, we can obtain such a flow distribution scheme within the same running-time. It remains to show that we can incorporate these results into a general procedure that yields a maximum flow within $\mathcal{O}(m)$ iterations.

In the following, for a feasible flow x, we refer to an edge e as *dead* if e is of type u or if there is a series-parallel subgraph G' that corresponds to an inner node in the decomposition tree, contains edge e, and satisfies $\alpha_x(G') = 0$. Note that a dead edge remains dead after an augmentation over any flow distribution scheme as described in the proof of Lemma 6.24.

Lemma 6.25:
For a flow x in a series-parallel graph G' with source s' that satisfies $\alpha_x(G') < 1$, let β denote the flow distribution scheme that is obtained by evaluating $\beta(G', \alpha_x(G'))$, where the function $\beta(\cdot, \cdot)$ is defined as in the proof of Lemma 6.24. It then holds that $\beta_e = \alpha_e$ for each $e \in \delta^+(s')$ that is of type α, and that $\beta_e = 0$ for each dead edge $e \in \delta^+(s')$. Moreover, there is no $e \in \delta^+(s')$ that is neither of type α nor dead.

Proof: If G' consists of a single edge $e = (s', t')$, the claim clearly follows according to the definition of β. Now let G' denote a series composition of two series-parallel subgraphs G_1 and G_2. If $\alpha_x(G') = 0$, it holds that all edges in G_1 and, thus, all edges in $\delta^+(s')$ are dead by definition. Since we obtain β_e for each $e \in \delta^+(s')$ by evaluating $\beta(G_1, \alpha_x(G')) = \beta(G_1, 0)$, the claim then follows. Otherwise, if $\alpha_x(G') \in (0, 1)$, it holds that $\alpha_x(G_1) = \alpha_x(G')$ by equation (6.4c) such that the values β_e for all edges in $\delta^+(s')$ are obtained by evaluating $\beta(G_1, \alpha_x(G_1))$ and the claim follows by induction. Finally, if G' is the parallel composition of two series-parallel subgraphs G_1 and G_2, it must hold that $\alpha_x(G_1) + \alpha_x(G_2) < 1$ and, thus, that $\alpha_x(G_1) < 1$ and $\alpha_x(G_2) < 1$. The claim then follows by induction since the values β_e for all $e \in \delta^+(s')$ are obtained by evaluating $\beta(G_1, \alpha_x(G_1))$ and $\beta(G_2, \alpha_x(G_2))$. $\qquad\square$

Lemma 6.26:
Let x denote a flow in a series-parallel graph G' that satisfies $\alpha_x(G') = 1$ and let β be an augmenting basic flow distribution scheme as described in the proof of Lemma 6.24. If $\beta_e > 0$ for some edge $e = (v, w)$, then, for each $e' \in \delta^+(v)$ with $e' \prec e$, it either holds that e' is dead and $\beta_{e'} = 0$ or that e' is of type α and $\beta_{e'} = \alpha_{e'}$.

Proof: Let e' and e with $e' \prec e$ be defined as above. Clearly, there is an inner node in the decomposition tree that corresponds to a parallel composition of two series-parallel subgraphs G_1 and G_2 with e' in G_1 and e in G_2. Let $0 < q \leqslant 1$ denote the fraction of additional flow that was distributed in two quantities $q_1 := \min\{q, \alpha_x(G_1)\}$ and $q_2 := q - q_1$ among G_1 and G_2, respectively, in the construction process of β as described in the proof of Lemma 6.24. Since $\beta_e > 0$, it holds that $q_2 > 0$ and, hence, that $\alpha_x(G_1) = q_1 < 1$. The claim then follows by Lemma 6.25. $\qquad\square$

Our augmenting flow distribution scheme algorithm for computing a maximum flow in a series-parallel generalized processing network works as follows: After computing the initial flow x as described in the proof of Lemma 6.21, we compute the value $\alpha_x(G)$ at the beginning of each iteration in $\mathcal{O}(m)$ time. If this value is less than one, we terminate and return the current flow x (we will show later that the flow is then optimum). Otherwise, we compute an augmenting basic flow distribution scheme β as described in the proof of Lemma 6.24. According to Lemma 6.11, we can determine the weight $w_\beta(e)$ of each edge $e \in E$ in $\mathcal{O}(m)$ time. The maximum amount of flow that can be sent on β is then given by the largest value $\delta > 0$ such that the flow $x(\delta)$ with $(x(\delta))_e := x_e + \delta \cdot w_\beta(e)$ is feasible. To be more precise, the value of δ is given by $\delta := \min\{\delta_1, \delta_2\}$ with

$$\delta_1 := \max\{\delta : (x(\delta))_e \leqslant u_e \ \forall e \in E\}$$
$$= \min\left\{ \frac{u_e - x_e}{w_\beta(e)} : e \in E \text{ with } w_\beta(e) > 0 \right\}.$$

and

$$\delta_2 := \max\left\{ \delta : (x(\delta))_e \leqslant \alpha_e \cdot \sum_{e' \in \delta^+(v)} (x(\delta))_{e'} \ \forall e = (v, w) \in E \right\}$$
$$= \max\left\{ \delta : x_e + \delta \cdot w_\beta(e) \leqslant \alpha_e \cdot \sum_{e' \in \delta^+(v)} (x_{e'} + \delta \cdot w_\beta(e')) \ \forall e = (v, w) \right\}$$

$$= \max \left\{ \delta : \delta \cdot \left(w_\beta(e) - \alpha_e \cdot \sum_{e' \in \delta^+(v)} w_\beta(e') \right) \leqslant \right.$$

$$\left. \left(\alpha_e \cdot \sum_{e' \in \delta^+(v)} x_{e'} \right) - x_e \ \forall e = (v,w) \in E \right\}$$

$$= \min \left\{ \frac{\left(\alpha_e \cdot \sum_{e' \in \delta^+(v)} x_{e'} \right) - x_e}{w_\beta(e) - \alpha_e \cdot \sum_{e' \in \delta^+(v)} w_\beta(e')} : e = (v,w) \in E \text{ with } \beta_e > \alpha_e \right\},$$

where the last equality follows from the fact that only those edges $e = (v,w) \in E$ restrict δ for which $w_\beta(e) - \alpha_e \cdot \sum_{e' \in \delta^+(v)} w_\beta(e') > 0$, which is true if and only if $\beta_e > \alpha_e$ since $w_\beta(e)$ is proportional to β_e and $\sum_{e' \in \delta^+(v)} \beta_{e'} = 1$.

Note that, for the value of δ determined above, at least one edge becomes of type α or of type u that was not of this type in x. We show in the following that at the same time it cannot happen that an edge that was dead or of type α before will be neither dead nor of type α after the augmentation. If we were not able to guarantee this, our procedure would not be guaranteed to terminate within a finite number of augmentations.

Lemma 6.27:
Let x' denote the flow that is obtained after augmenting a flow x with $\alpha_x(G) = 1$ over a flow distribution scheme β as described above and let $\mathcal{D}(x)$ $(\mathcal{D}(x'))$ and $\mathcal{A}(x)$ $(\mathcal{A}(x'))$ denote the set of dead edges and α-edges in x (x'), respectively. It then holds that $\mathcal{D}(x') \cup \mathcal{A}(x') \supsetneq \mathcal{D}(x) \cup \mathcal{A}(x)$ or that $\mathcal{D}(x') \cup \mathcal{A}(x') = \mathcal{D}(x) \cup \mathcal{A}(x)$ and $\mathcal{D}(x') \supsetneq \mathcal{D}(x)$.

Proof: First, consider some node $v \in V \setminus \{t\}$ and let $e \in \delta^+(v)$ denote the unique edge with $\beta_e > 0$ and $e' \prec e$ for each $e' \in \delta^+(v) \setminus \{e\}$ with $\beta_{e'} > 0$. According to Lemma 6.26, it holds that every such edge e' is either dead or of type α and remains so after an augmentation of value $\delta' < \delta$, where $\delta := \min\{\delta_1, \delta_2\}$ is defined as above. Moreover, if edge e was of type α as well before the augmentation, it holds that $\beta_e = 1 - \sum_{e' \in \delta^+(v): e' \prec e} \beta_{e'} = 1 - \sum_{e' \in \delta^+(v): e' \prec e, e' \in \mathcal{A}(x)} \alpha_{e'} = \alpha_e$ and e remains of type α. So, as long as we send less than δ units of flow on β, each edge remains its type.

By sending δ units of flow over β as described above, one of the following cases applies: If $\delta = \delta_1$, some edge becomes of type u in x' that was either of type α or of no type in x. In the first case, it holds that $\mathcal{D}(x') \cup \mathcal{A}(x') = \mathcal{D}(x) \cup \mathcal{A}(x)$ and $\mathcal{D}(x') \supsetneq \mathcal{D}(x)$, while in the second case we get that $\mathcal{D}(x') \cup \mathcal{A}(x') \supsetneq \mathcal{D}(x) \cup \mathcal{A}(x)$. On the other side, if $\delta = \delta_2$, some edge that was of no type in x becomes of type α in x', which implies that $\mathcal{D}(x') \cup \mathcal{A}(x') \supsetneq \mathcal{D}(x) \cup \mathcal{A}(x)$. This shows the claim. $\square$

Corollary 6.28:
The augmenting flow distribution scheme algorithm terminates after $\mathcal{O}(m)$ augmentations and runs in $\mathcal{O}(m^2)$ time.

Proof: The claim directly follows from the discussion before and from the combination of Lemma 6.27 and the fact that $|\mathcal{D}(x)| \leqslant m$ and $|\mathcal{A}(x)| \leqslant m$ for any flow x. $\square$

It remains to show that the computed flow is maximum. To this end, we need the following lemma:

Lemma 6.29:
Let G be a series-parallel graph and x be a flow in G that is positive on each edge $e \in E$. Moreover, assume that $\alpha_x(G) < 1$. Then there is an s-t-cut (S, T) such that

1. each edge in $\delta^+(S)$ is either of type α or of type u,

2. for each node $v \in V$ with $\emptyset \neq \delta^+(v) \subseteq \delta^+(S)$, it holds that at least one edge in $\delta^+(v)$ is of type u, and

3. $\delta^-(S) = \emptyset$.

Proof: Let x be a flow in a series-parallel graph G that fulfills $\alpha_x(G) < 1$. Consider the following function c_x defined on series-parallel subgraphs G' corresponding to nodes in the decomposition tree of G:

$$c_x(G' = e) = \{e\},$$
$$c_x(G' = G_1 \mid G_2) = c_x(G_1) \cup c_x(G_2),$$
$$c_x(G' = G_1 \circ G_2) = \begin{cases} c_x(G_2) & \text{if } \alpha_x(G_2) < 1, \\ c_x(G_1) & \text{else.} \end{cases}$$

We now show that the set $c_x(G)$ contains exactly the edges in an s-t-cut that fulfills the required properties. First, suppose that G consists of a single edge e. Since we are looking for an s-t-cut of G, the only possible cut is given by $\{e\} = c_x(G)$. Note that, since $\alpha_x(G) < 1$, edge e must be either of type u or of type α. Now assume that G is the series composition of two series-parallel graphs G_1 and G_2. Since $\alpha_x(G) < 1$, it holds that either $\alpha_x(G_2) < 1$ or that $\alpha_x(G_2) = 1$ but $\alpha_x(G_1) < 1$ (cf. equation (6.4c)). In the first (second) case, we set $c_x(G) := c_x(G_2)$ $(c_x(G) := c_x(G_1))$ and proceed recursively. Finally, suppose that G is the parallel composition of the two series-parallel graphs G_1 and G_2. Since each s-t-cut of G must pass both G_1 and G_2, we set $c_x(G) := c_x(G_1) \cup c_x(G_2)$.

Note that all the series-parallel subgraphs G' in the decomposition tree of G that contain edges in $c_x(G)$ fulfill $\alpha_x(G') < 1$. Hence, by evaluating $c_x(G)$ as described above, we obtain an s-t-cut (S,T) that only consists of edges of type α and type u, which shows claim (1). Now suppose that there is a node $v \in V$ with $\delta^+(v) \subseteq \delta^+(S)$ such that each edge in $\delta^+(v)$ is of type α. Since $x_e > 0$ for each $e \in E$, this implies that $\sum_{e \in \delta^+(v)} \alpha_e = 1$. Let G' denote the inclusionwise minimal series-parallel subgraph of G that contains all edges in $\delta^+(v)$ and corresponds to an inner node in the decomposition tree of G. For every series composition $G'' = G_1 \circ G_2$ that is contained in the decomposition tree of G', since $\delta^+(v) \subseteq \delta^+(S)$, it must hold that $\alpha_x(G_2) = 1$ and $\alpha_x(G_1) < 1$ and, thus, that $\alpha_x(G'') = \alpha_x(G_1)$. But then, according to the definition of c_x and α_x, it finally holds that $\alpha_x(G') = \sum_{e \in \delta^+(v)} \alpha_e = 1$, which contradicts the fact that $\alpha_x(G') < 1$ for all series-parallel subgraphs G' corresponding to an inner node in the decomposition tree of G.

Finally, the third claim follows from the fact that the graph G is acyclic and that, for each s-t-path P in G, the set $c_x(G)$ only contains one edge in P by construction. Hence, the cut that is implied by $c_x(G)$ fulfills all of the required properties and the claim of the lemma follows. $\qquad\square$

Lemma 6.30:
Let G be a series-parallel graph and let (S,T) denote an s-t-cut in G. Then there exists a node $v \in S$ with $\delta^+(v) \subseteq \delta^+(S)$.

Proof: Let $S' \subseteq S$ denote the set of all nodes in S that are reachable from s via a (possibly empty) directed path using nodes in S only. Since $s \in S'$, the set is nonempty. For a given topological sorting of the nodes in G, let $v \in S'$ denote the node in S' with the highest index in the topological sorting. Since $v \neq t$, it holds that $\delta^+(v) \neq \emptyset$. However, for each $e = (v,w) \in \delta^+(v)$, if e was not contained in $\delta^+(S)$, it would hold that $w \in S$ and, thus, that $w \in S'$. However, this would contradict the definition of v, which shows the claim. $\qquad\square$

We are now ready to prove the main theorem of this subsection:

Theorem 6.31:
The augmenting flow distribution scheme algorithm computes a maximum flow in a series-parallel generalized processing network in $\mathcal{O}(m^2)$ time.

Proof: The claimed running time follows from the above arguments. It remains to show that the computed flow x is maximum.

First, consider some arbitrary flow x' and the s-t-cut (S, T) obtained from applying Lemma 6.29 to the flow x computed by the algorithm. Using that $\text{excess}_{x'}(v) = 0$ for each $v \in V \setminus \{s, t\}$ according to equation (6.1b), we get that

$$\text{val}(x') = \text{excess}_x(t) = \sum_{v \in T} \text{excess}_{x'}(v) = \sum_{v \in T} \left(\sum_{e \in \delta^-(v)} x'_e - \sum_{e \in \delta^+(v)} x'_e \right)$$

$$= \sum_{v \in \delta^-(T)} x'_e - \sum_{v \in \delta^+(T)} x'_e = \sum_{v \in \delta^+(S)} x'_e - \sum_{v \in \delta^-(S)} x'_e \leqslant \sum_{v \in \delta^+(S)} x'_e,$$

i.e., the flow value of each flow x' is bounded by the total flow value on edges that head from S to T. We now show that the flow x that is computed by the augmenting flow distribution scheme algorithm fulfills $\text{val}(x) = \sum_{v \in \delta^+(S)} x_e$ and maximizes this value among all feasible flows, which shows the claim.

The first claim follows directly from the construction of the cut since there are no edges in $\delta^-(S)$. Now assume that there is a feasible flow x' with $\text{val}(x') > \text{val}(x)$. Clearly, there must be at least one edge in $\delta^+(S)$ for which $x'_e > x_e$. According to Lemma 6.30, there is at least one node $v \in S$ with $\delta^+(v) \subseteq \delta^+(S)$. Moreover, according to Lemma 6.29, it holds that each of the edges in $\delta^+(v)$ is of type α or of type u in x and that at least one of these edges is of type u, which, according to Lemma 6.20, implies that the flow leaving v is maximum and unique and, thus, the flow on each of the edges in $\delta^+(v)$ is maximum. Thus, it holds that $x'_e \leqslant x_e$ for each $e \in \delta^+(v)$. If $v = s$, it additionally holds that $\text{val}(x') = \sum_{e \in \delta^+(s)} x'_e \leqslant \sum_{e \in \delta^+(s)} x_e = \text{val}(x)$ and the claim of the theorem follows. Otherwise, v results from merging the sink of G_1 with the source of G_2 in a series composition of two series-parallel subgraphs G_1 and G_2. Let $\overline{G}$ denote the graph that results from G by replacing $G_1 \circ G_2$ with a single edge $\overline{e}$ with capacity $u_{\overline{e}} := \sum_{e \in \delta^+(v)} x_e$. Clearly, the flow $\overline{x}$ with $\overline{x}_{\overline{e}} := u_{\overline{e}}$ and $\overline{x}_e := x_e$ for each $e \in E(\overline{G}) \setminus \{\overline{e}\}$ is feasible in $\overline{G}$. Since there is a flow x' in G with $\text{val}(x') > \text{val}(x)$ and $x'_e \leqslant x_e$ for each $e \in \delta^+(v)$, there must be a flow $\overline{x}'$ with $\text{val}(\overline{x}') > \text{val}(\overline{x})$ in $\overline{G}$. However, since the new edge $\overline{e}$ is of type u in $\overline{x}$ and is contained in the cut $(\overline{S}, \overline{T})$ with $\overline{S} := S \cap V(\overline{G})$ and $\overline{T} := T \cap V(\overline{G})$, we can repeat the above arguments until $S = \{s\}$ and $\delta^+(s) = \delta^+(S)$. Thus, there is no flow x' in G with $\text{val}(x') > \text{val}(x)$ and the claim follows. $\square$

6.5.2 A Faster Approach

Lemma 6.18 and Corollary 6.19 already give insights about the behavior of a special case of series-parallel graphs, namely the case of parallel edges between two nodes v and w. We now show how to incorporate the behavior of other edges in $\delta^+(v)$ that do not reach w into the above algorithm. Until the end of the following discussion, we again assume that the edges in $E(v, w) = \{e_1, \ldots, e_k\}$ are ordered such that $\frac{u_{e_i}}{\alpha_{e_i}} \leqslant \frac{u_{e_j}}{\alpha_{e_j}}$ for $i < j$.

Consider the set $E(v, w) := \{e_1, \ldots, e_k\}$ of parallel edges between two nodes v and w and assume that the total flow on the remaining edges $e \in \delta^+(v) \setminus E(v, w)$ that leave v is given by x_0. For a fixed value of x_0, we can find the maximum flow on the edges in $E(v, w)$ by using the algorithm described in the proof of Corollary 6.19 as follows: By introducing an artificial edge e_0 between v and w with capacity $u_{e_0} := x_0$ and $\alpha_{e_0} := 1$ and evaluating the algorithm on the graph $G' := (\{v, w\}, E(v, w) \cup \{e_0\})$, edge e_0 will eventually be labeled as type u since $\alpha_{e_0} = 1$ (so the second case must hold in the proof of Lemma 6.18 when we reach edge e_0) and will, thus, carry the desired amount of flow x_0 while $\sum_{e \in E(v, w)} x_e$ is maximum according to Corollary 6.19.

Now let x_0 be of variable value. Note that, for each $e_i \in E(v, w)$, there is some (not necessarily positive) value $b(e_i)$ such that edge e_i is declared to be of type α for $x_0 < b(e_i)$ and is labeled as type u for $x_0 \geqslant b(e_i)$. We call this value $b(e_i)$ the *breakpoint of edge* e_i in the following. Clearly, according to the proof of Lemma 6.18, it holds that

$$x_0 = b(e_i) \quad \Longleftrightarrow \quad \alpha_{e_i} = \left(1 - \alpha^{(i)}\right) \cdot \frac{u_{e_i}}{x_0 + \sum_{j=1}^{i} u_{e_j}}$$

$$\Longleftrightarrow \quad x_0 = \left(1 - \alpha^{(i)}\right) \cdot \frac{u_{e_i}}{\alpha_{e_i}} - \sum_{j=1}^{i} u_{e_j} \qquad (6.5)$$

with $\alpha^{(i)} = \sum_{j=i+1}^{k} \alpha_{e_j}$ as before. Using the sorting of the edges in $E(v, w)$, we get the following result:

Lemma 6.32:
Consider a set of edges $E(v, w) = \{e_1, \ldots, e_k\}$ with $\frac{u_{e_i}}{\alpha_{e_i}} \leqslant \frac{u_{e_j}}{\alpha_{e_j}}$ for $i < j$ between two nodes $v, w \in V$. For $i \in \{1, \ldots, k-1\}$, it holds that $b(e_i) \leqslant b(e_{i+1})$.

Proof: Using the definitions of $b(e_i)$ and $b(e_{i+1})$ and the ordering of the edges in $E(v, w)$, we get that

$$b(e_i) = \left(1 - \alpha^{(i)}\right) \cdot \frac{u_{e_i}}{\alpha_{e_i}} - \sum_{j=1}^{i} u_{e_j} \leqslant \left(1 - \alpha^{(i)}\right) \cdot \frac{u_{e_{i+1}}}{\alpha_{e_{i+1}}} - \sum_{j=1}^{i} u_{e_j}$$

$$= \left(1 - \alpha_{e_{i+1}} - \alpha^{(i+1)}\right) \cdot \frac{u_{e_{i+1}}}{\alpha_{e_{i+1}}} - \sum_{j=1}^{i} u_{e_j}$$

$$= \left(1 - \alpha^{(i+1)}\right) \cdot \frac{u_{e_{i+1}}}{\alpha_{e_{i+1}}} - u_{e_{i+1}} - \sum_{j=1}^{i} u_{e_j}$$

$$= \left(1 - \alpha^{(i+1)}\right) \cdot \frac{u_{e_{i+1}}}{\alpha_{e_{i+1}}} - \sum_{j=1}^{i+1} u_{e_j} = b(e_{i+1}),$$

which shows the claim. □

Lemma 6.33:

Let $f_{(v,w)}(x_0)$ denote the maximum flow that can be sent through the parallel edges $E(v,w) = \{e_1, \ldots, e_k\}$ depending on the total flow x_0 on the remaining edges in $\delta^+(v) \setminus E(v,w)$. The function $f_{(v,w)}$ is continuous, non-decreasing, concave, and piecewise linear with breakpoints contained in $\{b(e_i) : i \in \{1, \ldots, k\}\}$ and non-negative slopes between two adjacent breakpoints. Moreover, it holds that $f_{(v,w)}(x_0) > 0$ for each $x_0 \geqslant 0$. The function can be determined in $\mathcal{O}(k)$ time.

Proof: First consider the case that $x_0 = 0$. According to Lemma 6.18 and 6.20, there is some index h such that, in every maximum flow x between v and w, we have $x_{e_i} = u_{e_i}$ for $1 \leqslant i \leqslant h$ and $x_{e_j} = \alpha_{e_j} \cdot F$ for $h+1 \leqslant j \leqslant k$, where $F = x_0 + \sum_{i=1}^{k} x_{e_i}$ is the total outflow of node v. If $h = k$, the claim clearly follows since, in this case, $f_{(v,w)}$ is a constant function. Else, the partitioning of the edges into the types u and α remains valid until the first breakpoint is reached. According to Lemma 6.32, this breakpoint is given by $b(e_{h+1})$. Note that, for $x_0 \in [0, b(e_{h+1}))$, the maximum flow through the edges in $E(v,w)$ is given by

$$f_{(v,w)}(x_0) = \sum_{i=1}^{k} x_{e_i} = \sum_{i=1}^{h} u_{e_i} + \sum_{j=h+1}^{k} \alpha_{e_j} \cdot F$$

$$= \sum_{i=1}^{h} u_{e_i} + \left(f_{(v,w)}(x_0) + x_0\right) \cdot \alpha^{(h)},$$

so

$$f_{(v,w)}(x_0) = \frac{1}{1 - \alpha^{(h)}} \cdot \left(\sum_{i=1}^{h} u_{e_i} + x_0 \cdot \alpha^{(h)}\right) > 0, \tag{6.6}$$

which is an increasing linear function of x_0 with slope $\frac{\alpha^{(h)}}{1-\alpha^{(h)}} > 0$. Again, if $h+1 = k$, the claim follows.

Else, as x_0 reaches the value $b(e_{h+1})$, edge e_{h+1} turns from type α to type u and, accordingly, the function $f_{(v,w)}(x_0)$ behaves as a linear function of x_0 until the next breakpoint $b(e_{h+2})$ is reached and so on. The slope of $f_{(v,w)}$ on $[b(e_{h+1}), b(e_{h+2}))$ evaluates to $\frac{\alpha^{(h+1)}}{1-\alpha^{(h+1)}} \leqslant \frac{\alpha^{(h)}}{1-\alpha^{(h)}}$, so the slopes of the linear segments do not increase with x_0.

It remains to show the continuity of $f_{(v,w)}$, which – in combination with the above arguments – in turn yields that the function is concave and non-decreasing. Consider the intersection of the above two adjacent linear segments of $f_{(v,w)}$:

$$\frac{1}{1-\alpha^{(h)}} \cdot \left(\sum_{i=1}^{h} u_{e_i} + x_0 \cdot \alpha^{(h)} \right) = \frac{1}{1-\alpha^{(h+1)}} \cdot \left(\sum_{i=1}^{h+1} u_{e_i} + x_0 \cdot \alpha^{(h+1)} \right).$$

Multiplying by $(1-\alpha^{(h)})$ and $(1-\alpha^{(h+1)})$, we get that

$$\left(1 - \alpha^{(h+1)} \right) \cdot \left(\sum_{i=1}^{h} u_{e_i} + x_0 \cdot \alpha^{(h)} \right) = \left(1 - \alpha^{(h)} \right) \cdot \left(\sum_{i=1}^{h+1} u_{e_i} + x_0 \cdot \alpha^{(h+1)} \right). \quad (6.7)$$

The left-hand side of equation (6.7) evaluates to

$$\left(1 - \alpha^{(h+1)} \right) \cdot \left(\sum_{i=1}^{h} u_{e_i} + x_0 \cdot \alpha^{(h+1)} \right) + \left(1 - \alpha^{(h+1)} \right) \cdot x_0 \cdot \alpha_{e_{h+1}},$$

while the right-hand side of equation (6.7) can be rearranged into

$$\left(1 - \alpha^{(h+1)} \right) \cdot \left(\sum_{i=1}^{h+1} u_{e_i} + x_0 \cdot \alpha^{(h+1)} \right) - \alpha_{e_{h+1}} \cdot \left(\sum_{i=1}^{h+1} u_{e_i} + x_0 \cdot \alpha^{(h+1)} \right).$$

Hence, subtracting $\left(1 - \alpha^{(h+1)} \right) \cdot \left(\sum_{i=1}^{h} u_{e_i} + x_0 \cdot \alpha^{(h+1)} \right)$ from both sides of equation (6.7) and dividing by $\alpha_{e_{h+1}}$ yields

$$\left(1 - \alpha^{(h+1)} \right) \cdot x_0 = \left(1 - \alpha^{(h+1)} \right) \cdot \frac{u_{e_{h+1}}}{\alpha_{e_{h+1}}} - \left(\sum_{i=1}^{h+1} u_{e_i} + x_0 \cdot \alpha^{(h+1)} \right)$$

$$\Longleftrightarrow \qquad x_0 = \left(1 - \alpha^{(h+1)} \right) \cdot \frac{u_{e_{h+1}}}{\alpha_{e_{(h+1)}}} - \sum_{i=1}^{h+1} u_{e_i},$$

i.e., both line segments intersect at $x_0 = b(e_{h+1})$. Repeating the above arguments until the last breakpoint is reached then yields the claim. $\qquad\square$

Lemma 6.34:
Consider three nodes $v, w, z \in V$ with parallel edges $E(v, w) = \{e_1, \ldots, e_k\}$ between v and w and $\delta^+(w) = \{\bar{e}\}$, where $\bar{e}$ is heading to z. The set of edges $E(v, w) \cup \{\bar{e}\}$ can be replaced by $i \leqslant k$ parallel edges $\{e_1', \ldots, e_i'\}$ from v to z such that, for each flow value x_0 on the edges in $\delta^+(v) \setminus E(v, w)$, the maximum amount of flow that can arrive at z through the edges $\{e_1', \ldots, e_i'\}$ equals the maximum amount of flow that can reach z through $E(v, w)$ and $\bar{e}$. This transformation can be performed in $\mathcal{O}(k)$ time.

Proof: Let $g_{(v,z)}(x_0)$ denote the maximum amount of flow that can be sent from v to z using the edges in $E(v, w)$ and the edge $\bar{e}$ depending on the total flow value x_0

on the edges in $\delta^+(v) \setminus E(v,w)$. Clearly, due to the structure of the subgraph that is considered, $g_{(v,z)}(x_0) = \min\{u_{\bar{e}}, f_{(v,w)}(x_0)\}$, where $f_{(v,w)}$ is defined as in Lemma 6.33. If $u_{\bar{e}} \geqslant \max_{x_0 \geqslant 0} f_{(v,w)}(x_0)$, the claim follows by deleting the edge $\bar{e}$ and merging the nodes w and z.

Else, if $u_{\bar{e}} < \max_{x_0 \geqslant 0} f_{(v,w)}(x_0)$, it either holds that $u_{\bar{e}} < f_{(v,w)}(0)$ or there must be two adjacent breakpoints $b(e_{i-1})$ and $b(e_i)$ with $f_{(v,w)}(b(e_{i-1})) \leqslant u_{\bar{e}} < f_{(v,w)}(b(e_i))$ (which are uniquely defined since $f_{(v,w)}$ is non-decreasing and continuous according to Lemma 6.33). In the first case, it clearly holds that $g_{(v,z)}(x_0) = u_{\bar{e}}$ and the claim follows by deleting the edges in $E(v,w)$ and merging the nodes v and w.

Now assume that $f_{(v,w)}(b(e_{i-1})) \leqslant u_{\bar{e}} < f_{(v,w)}(b(e_i))$ for some $i \in \{2,\ldots,k\}$ and let $\bar{x}$ denote the (uniquely defined) flow value such that $f_{(v,w)}(\bar{x}) = u_{\bar{e}}$. Note that all of the edges e_j with $i \leqslant j \leqslant k$ remain of type α within the interval $[0,\bar{x}]$ since the respective breakpoints are not yet reached, i.e., a constant fraction $\alpha^{(i-1)} = \sum_{j=i}^{k} \alpha_{e_j}$ of $f_{(v,w)}(x_0) = g_{(v,z)}(x_0)$ flows through the edges $e_i,\ldots,e_k$ as long as $x_0 \in [0,\bar{x}]$. Hence, we can replace the edges $e_i,\ldots,e_k$ by a single edge e_i' with $\alpha_{e_i'} := \alpha^{(i-1)}$ without changing the behavior of $f_{(v,w)}$ and $g_{(v,z)}$ in $[0,\bar{x}]$. Since e_i' is the only edge of type α within the interval $[b(e_{i-1}),\bar{x}]$, we achieve that $f_{(v,w)}(x_0) = u_{\bar{e}} = g_{(v,z)}(x_0)$ for $x_0 \geqslant \bar{x}$ by setting $u_{e_i'} := u_{\bar{e}} - \sum_{j=1}^{i-1} u_{e_j}$.

Thus, after the transformation, it holds that $f_{(v,w)}(x_0) = g_{(v,z)}(x_0)$ for each $x_0 \geqslant 0$, i.e., edge $\bar{e}$ does not influence the flow value anymore. By deleting the edge $\bar{e}$ and merging the nodes w and z, the claim then follows. $\qquad\square$

Note that the edges in $E'(v,w) := \{e_1',\ldots,e_l'\}$ remain ordered by $\frac{u_{e_j'}}{\alpha_{e_j'}}$ after the transformation performed in the proof of Lemma 6.34: For the case that $i = 1$, the claim clearly holds. Otherwise, by construction, it holds that $u_{\bar{e}} \geqslant f_{(v,w)}(b(e_{i-1}))$, so

$$
\frac{u_{e_i'}}{\alpha_{e_i'}} = \frac{u_{\bar{e}} - \sum_{j=1}^{i-1} u_{e_j}}{\alpha^{(i-1)}} \geqslant \frac{f_{(v,w)}(b(e_{i-1})) - \sum_{j=1}^{i-1} u_{e_j}}{\alpha^{(i-1)}}
$$

$$
\overset{(6.6)}{=} \frac{\frac{1}{1-\alpha^{(i-1)}} \cdot \left(\sum_{j=1}^{i-1} u_{e_j} + b(e_{i-1}) \cdot \alpha^{(i-1)}\right) - \sum_{j=1}^{i-1} u_{e_j}}{\alpha^{(i-1)}}
$$

$$
= \frac{\left(\frac{1}{1-\alpha^{(i-1)}} - 1\right) \cdot \sum_{j=1}^{i-1} u_{e_j} + b(e_{i-1}) \cdot \frac{\alpha^{(i-1)}}{1-\alpha^{(i-1)}}}{\alpha^{(i-1)}}
$$

$$
= \frac{\frac{\alpha^{(i-1)}}{1-\alpha^{(i-1)}} \cdot \sum_{j=1}^{i-1} u_{e_j} + b(e_{i-1}) \cdot \frac{\alpha^{(i-1)}}{1-\alpha^{(i-1)}}}{\alpha^{(i-1)}} = \frac{\sum_{j=1}^{i-1} u_{e_j} + b(e_{i-1})}{1-\alpha^{(i-1)}}
$$

$$
\overset{(6.5)}{=} \frac{\sum_{j=1}^{i-1} u_{e_j} + \left(1 - \alpha^{(i-1)}\right) \cdot \frac{u_{e_{i-1}}}{\alpha_{e_{i-1}}} - \sum_{j=1}^{i-1} u_{e_j}}{1-\alpha^{(i-1)}} = \frac{u_{e_{i-1}}}{\alpha_{e_{i-1}}},
$$

where the last inequality follows from Lemma 6.32. Since the transformation assures that $e'_j = e_j$ for $j \in \{1, \ldots, i-1\}$, we get the following observation:

Observation 6.35:
After applying the procedure described in Lemma 6.34 to a set of edges $E(v, w) = \{e_1, \ldots, e_k\}$ and $\bar{e} = (w, z)$, the resulting edge set $\{e'_1, \ldots, e'_i\}$ is ordered by the values $\frac{u_{e'_j}}{\alpha_{e'_j}}$ again. ◁

Lemma 6.36:
Let v and w be two nodes such that all edges $\{e_1, \ldots, e_k\}$ that leave v are parallel edges heading to w and let F denote the maximum flow that can be sent from v to w. Then each flow value F' with $0 \leqslant F' \leqslant F$ can be achieved as well.

Proof: The claim follows by a simple scaling argument: Consider the maximum flow x on the edges $e_1, \ldots, e_k$ with flow value F that is, e.g., determined using the algorithm described in the proof of Corollary 6.19. It is easy to see that the flow x' with $x'_{e_i} := \frac{F'}{F} \cdot x_{e_i}$ for each $i \in \{1, \ldots, k\}$ is feasible as well and achieves a flow value of F'. □

Theorem 6.37:
A maximum flow in a series-parallel graph $G = (V, E)$ can be computed in $\mathcal{O}(m \cdot (n + \log m))$ time.

Proof: Consider a decomposition tree T of G. By sorting the leaves of T in $\mathcal{O}(m \cdot \log m)$ time, we can assure that $\frac{u_{e_i}}{\alpha_{e_i}} \leqslant \frac{u_{e_j}}{\alpha_{e_j}}$ for $i < j$ in each set of edges $\{e_1, \ldots, e_k\}$.

In a second step, we use a breadth-first-search in the decomposition tree starting at the root in order to get a list S of all nodes that correspond to series compositions in $\mathcal{O}(m)$ time. Note that this list is inherently sorted by the depth of the respective nodes in the tree.

Let $v \in T$ denote a node in T of maximum depth that corresponds to a series-parallel graph G' that is the series composition of two series-parallel graphs G_1 and G_2 (note that v can be found in $\mathcal{O}(1)$ time by looking at the tail of S). Due to the maximum depth of v, neither G_1 nor G_2 can contain series compositions, i.e., each of the graphs either consists of a single edge or of parallel edges.

Let k_1 and k_2 denote the number of edges contained in G_1 and G_2, respectively. First, consider the case that G_2 consists of parallel-edges $e_1, \ldots, e_{k_2}$ with $k_2 \geqslant 2$. Due to the structure of series-parallel graphs, there are no other edges leaving the source node of G_2, so we can find the maximum amount of flow F_2 that can be sent through G_2 in $\mathcal{O}(k_2)$ time according to Corollary 6.19. Consequently, we can replace the edges in G_2 by a single edge with capacity F_2. This in turn enables us to replace the edges in G_1

and G_2 by at most k_1 edges E' in $\mathcal{O}(k_1)$ time using Lemma 6.34. Note that these edges are ordered according to Observation 6.35, but the set $\delta^+(s_{G'})$, where $s_{G'}$ is the source of G', may not be ordered anymore. However, since both the edges in E' and the edges in $\delta^+(s_{G'}) \setminus E'$ are sorted, we can regain the ordering of $\delta^+(s_{G'})$ in $\mathcal{O}(m)$ time.

The algorithm stops when no series composition is left, i.e., the remaining graph consists of parallel edges only. The maximum flow value can then be determined using the procedure described in the proof of Corollary 6.19. Since the maximum flow in each series composition can be computed in $\mathcal{O}(m)$ time as described above, the claimed running time follows. Note that the procedure only describes the computation of the maximum flow value in G. However, using the procedures described in the proofs of Corollary 6.19, Lemma 6.34, and Lemma 6.36, the flow on each edge can be determined as well within the same running time. □

6.6 Integral Flows

In the traditional maximum flow problem, there always exists an integral optimal solution if the capacities are integral (cf. Section 2.4). It is easy to see that this is no longer valid for the case of flows in processing networks. In particular, if we add the requirement that the flow on each edge needs to be integral, the maximum flow problem in a generalized processing network becomes both $\mathcal{NP}$-hard to solve and to approximate as we will see in the following two theorems.

Theorem 6.38:
The problem of finding a maximum *integral* flow in a generalized processing network is strongly $\mathcal{NP}$-complete to solve and $\mathcal{NP}$-hard to approximate within constant factors, even if the graph is acyclic and bipartite.

Proof: We first show the $\mathcal{NP}$-completeness of the problem. Clearly, the problem is contained in $\mathcal{NP}$ since we can check if a given solution candidate (which has a polynomially bounded encoding length) is feasible and has a specific flow value in polynomial time. In order to show $\mathcal{NP}$-hardness, we use a reduction from the EXACT-COVERBY3SETS-problem, which is known to be strongly $\mathcal{NP}$-complete (cf. (Garey and Johnson, 1979, Problem SP2)):

INSTANCE: A set X with 3q elements and a collection $\mathcal{C} = \{C_1, \ldots, C_k\}$ of 3-element subsets of X.

QUESTION: Does there exist a subcollection $\mathcal{C}' \subseteq \mathcal{C}$ such that every element $j \in X$ is contained in exactly one of the subsets in $\mathcal{C}'$?

Given an instance of ExactCoverBy3Sets, we construct a generalized processing network as follows:

We introduce a source s and a sink t as well as nodes v_i and v_i' for each $C_i \in \mathcal{C}$ and a node w_j for each $j \in X$. For each subset $C_i \in \mathcal{C}$, we insert an edge with capacity 3 between s and v_i and three edges, each with flow ratio of $\frac{1}{3}$ between v_i and v_i'. Furthermore, we introduce an edge between v_i' and w_j if $j \in C_i$ and an edge between each node w_j and the sink t. If not mentioned explicitly, we set $\alpha_e = 1$ and $u_e = 1$ for every other edge $e \in E$. The resulting network for $X = \{1, \ldots, 9\}$ and $\mathcal{C} = \{\{1,2,4\}, \{2,3,4\}, \{3,5,8\}, \{4,6,7\}, \{6,7,9\}\}$ is shown in Figure 6.2. It is easy to see that the constructed network is always acyclic and bipartite, as claimed.

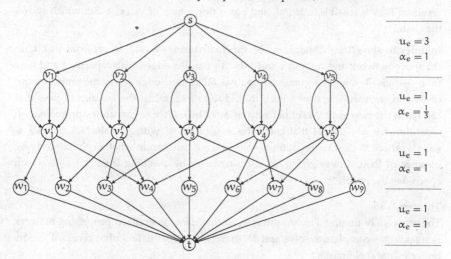

$$u_e = 3$$
$$\alpha_e = 1$$

$$u_e = 1$$
$$\alpha_e = \frac{1}{3}$$

$$u_e = 1$$
$$\alpha_e = 1$$

$$u_e = 1$$
$$\alpha_e = 1$$

Figure 6.2: The resulting network for a given instance of ExactCoverBy3Sets with $X = \{1, \ldots, 9\}$ and $\mathcal{C} = \{\{1,2,4\}, \{2,3,4\}, \{3,5,8\}, \{4,6,7\}, \{6,7,9\}\}$. On the right hand side, the capacities and flow ratios of the edges in each level of the graph are depicted.

We now show that there exists an integral flow x with flow value $\mathrm{val}(x) \geqslant 3q$ if and only if the underlying instance of ExactCoverBy3Sets is a Yes-instance. First assume that there is an integral flow x with flow value $\mathrm{val}(x) \geqslant 3q$. In fact, since there are $3q$ edges leading to the sink, each with a capacity of one, it must hold that $\mathrm{val}(x) = 3q$. Thus, each of these edges must carry one unit of flow. Moreover, note that each of the edges between s and v_i for $i \in \{1, \ldots, k\}$ may either carry zero or three units of flow since there are three edges between each v_i and v_i' that are linked such that they can only carry the same (integral) amount of flow at the same time. Hence, the amount of flow that arrives at the nodes $v_1', \ldots, v_k'$ equals three units for exactly q of these nodes and zero for the remaining $k - q$ nodes. Let $K \subseteq \{1, \ldots, k\}$ denote the set of the indices of those nodes v_i' for which three units of flow arrive. Since each of the $3q$ edges between w_j and t, for $j \in \{1, \ldots, 3q\}$, carries one unit of flow as described above, it, thus, follows that the adjacent nodes to the nodes v_i' for $i \in K$ are disjoint, which in turn implies that the corresponding sets C_i are mutually disjoint and cover X. Hence, $\mathcal{C}' := \{C_i : i \in K\}$ is a solution to the underlying instance of ExactCoverBy3Sets.

Now suppose that $\mathcal{C}'$ is a solution to a given instance of ExactCoverBy3Sets. For each $C_i \in \mathcal{C}'$, we send three units of flow from s to v_i and one unit on each of the three edges between v_i and v_i'. Since the sets in $\mathcal{C}'$ do not intersect, we can send one unit of flow to each of the nodes w_j for $j \in \{1, \ldots, 3q\}$ and finally to the sink. The resulting flow is feasible, integral, and has a flow value of $\mathrm{val}(x) = 3q$, which shows the claim.

In order to show the $\mathcal{NP}$-hardness of approximation, we add an artificial sink t' to the above network and connect t and t' by $3q$ parallel edges with capacity 1 and flow ratio $\frac{1}{3q}$ each. By similar arguments as above, all of these edges carry the same amount of flow in each feasible flow x such that, due to integrality, the amount of flow that can reach the new sink t' is either zero or $3q$. Thus, if there was an α-approximation algorithm for $\alpha \in (1, \infty)$ that computes a solution x' with $\frac{1}{\alpha} \cdot \mathrm{val}(x^*) \leqslant \mathrm{val}(x') \leqslant \mathrm{val}(x^*)$ where x^* denotes an optimal solution, we can decide whether the underlying instance of ExactCoverBy3Sets is a Yes-instance by checking if $\mathrm{val}(x') > 0$, which concludes the proof. $\qquad\Box$

Theorem 6.39:

The problem of finding a maximum *integral* flow in a generalized processing network is weakly $\mathcal{NP}$-complete to solve and $\mathcal{NP}$-hard to approximate within constant factors on series-parallel graphs.

Proof: For the reduction, we use the SubsetSum-problem, which is defined as follows (cf. (Garey and Johnson, 1979, Problem SP13)):

INSTANCE: Finite set $\{a_1, \ldots, a_k\}$ of k positive integers and a positive integer A.

QUESTION: Is there a subset $I \subseteq \{1, \ldots, k\}$ such that $\sum_{i \in I} a_i = A$?

Given an instance of SubsetSum, we construct a series-parallel generalized processing network as follows:

We insert three nodes s, v_0, and t. Between v_0 and t, we introduce two parallel edges, one of them with capacity 1 and flow ratio $\frac{1}{A}$ and the other one with capacity $A - 1$ and flow ratio $\frac{A-1}{A}$. Moreover, for each $i \in \{1, \ldots, k\}$, we insert an additional node v_i, an edge between s and v_i with capacity a_i and flow ratio 1, and two parallel edges between v_i and v_0, one of them with capacity 1 and flow ratio $\frac{1}{a_i}$ and the other one with capacity $a_i - 1$ and flow ratio $\frac{a_i-1}{a_i}$. The resulting graph is depicted in Figure 6.3.

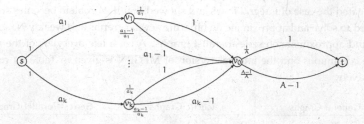

Figure 6.3: The resulting network for a given instance of SUBSETSUM. The edges are labeled with the corresponding capacities while the edge tails are labeled with the flow ratios.

Note that, due to integrality, the flow that reaches the sink is either zero or A, since the flow on the edge with capacity 1, which amounts to either zero or one, only makes up a fraction $\frac{1}{A}$ of the flow that reaches t. Accordingly, for $i \in \{1, \ldots, k\}$, the amount of flow that reaches v_0 via v_i may either be zero or a_i. By identifying those nodes v_i that receive a_i units of flow from s with those elements that are contained in I, the claims follow by similar arguments as in the proof of Theorem 6.38. □

6.7 Conclusion

In this chapter, we generalized the well-known maximum flow problem in processing networks from flow ratios that determine the *exact* fraction of flow routed through an edge to flow ratios that determine an *upper bound* on this fraction. We were able to generalize the notion of paths as a central unit in traditional network flows to the concept of flow distribution schemes and could show that a flow decomposition similar to the one for traditional network flows is possible on general graphs. Although it was easy to see that the problem is solvable in weakly polynomial time, we could show that the problem is at least as hard to solve as any packing LP. Nevertheless, for acyclic graphs, we presented a (strongly polynomial-time) FPTAS with a running time of $\mathcal{O}\left(\frac{1}{\varepsilon^2} \cdot m^2 \log m\right)$ that also embodies the first approximation scheme for the maximum flow problem in processing networks. Moreover, for the case of series-parallel graphs, we presented two very different approaches on how to solve the problem exactly. The first of these approaches generalizes the idea of the augmenting path algorithm for the traditional maximum flow problem and resulted in an algorithm with a running time of $\mathcal{O}(m^2)$. The second approach used the structure of series-parallel graphs in a more sophisticated way in order to improve this running time based on a successive shrinking of subgraphs to $\mathcal{O}(m \cdot (n + \log m))$. Finally, we investigated the case of integral flows and showed that the problem becomes strongly $\mathcal{NP}$-hard to solve and approximate on bipartite acyclic graphs and weakly $\mathcal{NP}$-hard to solve and approximate on series-parallel graphs. A complete overview of the results for the continuous and the integral version of MFGPN is given in Table 6.1 and 6.2, respectively.

General Graphs	Acyclic Graphs	Series-Parallel Graphs
Theorem 6.10: Decomposable in $\mathcal{O}(m^4)$ time	Theorem 6.13: Decomposable in $\mathcal{O}(m^2)$ time	$\longrightarrow$
$\longleftarrow$	Theorem 6.15: At least as hard to solve as any packing LP	
Theorem 6.14: Solvable in $\mathcal{O}(m^{3.5} \log M)$ time	$\longrightarrow$	Theorem 6.31: Solvable in $\mathcal{O}(m^2)$ time Theorem 6.37: Solvable in $\mathcal{O}(m \cdot (n + \log m))$ time
	Theorem 6.17: FPTAS in $\mathcal{O}\left(\frac{1}{\epsilon^2} m^2 \log m\right)$ time	$\longrightarrow$

Table 6.1: The summarized results for the continuous maximum flow problem in generalized processing networks in Chapter 6. Implied results are denotes with gray arrows.

General Graphs	Acyclic Graphs	Series-Parallel Graphs
$\longleftarrow$	Theorem 6.38: Strongly $\mathcal{NP}$-complete to solve	Theorem 6.39: Weakly $\mathcal{NP}$-complete to solve
$\longleftarrow$	Theorem 6.38: $\mathcal{NP}$-hard to approximate	Theorem 6.39: $\mathcal{NP}$-hard to approximate

Table 6.2: The summarized results for the integral maximum flow problem in generalized processing networks in Chapter 6. Implied results are denoted with gray arrows.

The introduced generalized model provides many topics for future research. On the one hand, it would be worth investigating if a network simplex algorithm as described in (Wang and Lin, 2009) could be adopted to the extended model. Although the problem lies in $\mathcal{P}$, there are no polynomial-time combinatorial algorithms for the maximum flow problem both in the traditional and the generalized model yet. On the other hand, the introduced model could be further extended to the case of minimum cost flows. It remains open if some of the algorithms introduced in this chapter can be generalized in order to compute a minimum cost flow in a given acyclic or series-parallel network.

7 | Convex Generalized Flows

In this chapter, we give insights into the structural properties and the complexity of an extension of the generalized maximum flow problem in which the outflow of an edge is a strictly increasing convex function of its inflow. In contrast to the traditional generalized maximum flow problem, which is solvable in polynomial time as shown in Section 2.4, we show that the problem becomes $\mathcal{NP}$-hard to solve and approximate in this novel setting. Nevertheless, we show that a flow decomposition similar to the one for traditional generalized flows is possible and present (exponential-time) exact algorithms for computing optimal flows on acyclic, series-parallel, and extension-parallel graphs as well as optimal preflows on general graphs. We also identify a polynomially solvable special case and show that the problem is solvable in pseudo-polynomial time when restricting to integral flows on series-parallel graphs.

This chapter is based on joint work with Sven O. Krumke and Clemens Thielen (Holzhauser et al., 2015b).

7.1 Introduction

As it was shown in Section 2.4, the traditional generalized flow problem may be used in order to model real world scenarios such as the loss of water in a broken pipe or the conversion of money between currencies. However, the fixed ratio between the outflow and the inflow of an edge that comes with traditional generalized flows may be insufficient in several applications. In this chapter, we investigate an extension of generalized flows from linear outflow functions $g_e(x_e) = \gamma_e \cdot x_e$ to general strictly increasing continuous convex outflow functions g_e. These more general outflow functions enable us to model processes in which the effectiveness increases with the load. This happens, e.g., in various trading applications where better rates are obtained if larger amounts are traded. By identifying the possible goods with nodes and introducing edges to represent trading options, this effect can be modeled more realistically with the help of convex outflow functions. For two nodes representing the goods A and B, a flow of maximum flow value between these two nodes then yields a strategy for obtaining the maximum amount of good B out of a existing stock of goods of type A.

7.1.1 Previous Work

The traditional generalized maximum flow problem with linear outflow functions has been studied extensively in the past fifty years and is still a topic of active research. In 1977, Truemper (1977) discovered an analogy between generalized maximum flows and traditional minimum cost flows and noted that many of the combinatorial algorithms for the generalized maximum flow problem known at this time were in fact pseudo-polynomial time variations of well-known algorithms for standard minimum cost flow problems. Although the generalized maximum flow problem can be expressed as a linear program and solved in polynomial time, e.g., by interior point methods, it took until 1991 that Goldberg et al. (1991) developed the first (weakly) polynomial-time combinatorial algorithm for the generalized maximum flow problem. For series-parallel graphs, a strongly polynomial-time algorithm was presented by Krumke and Zeck (2013). The first strongly polynomial-time algorithm for general graphs was recently given by Végh (2013).

Ahlfeld et al. (1987) and Tseng and Bertsekas (2000) studied extensions of generalized flows in which the objective function is replaced by a nonlinear function and a minimum cost flow is sought. Nevertheless, the outflow functions are assumed to be linear in both papers. Nonlinear outflow functions in the generalized maximum flow problem have first been studied by Truemper (1978) and later by Shigeno (2006). In both papers, a generalization to (increasing and continuous) *concave* outflow functions is suggested and optimality criteria are presented. By exploiting several analogies to the case of linear outflow functions, Végh (2012) obtained an efficient combinatorial algorithm for this problem. This algorithm, however, makes heavy use of the concavity of the outflow functions, so it cannot be used for the case of convex outflow functions studied here. To the best of our knowledge, (general) convex outflow functions in the generalized maximum flow problem have not been studied so far.

7.1.2 Chapter Outline

After the introduction of the necessary assumptions and definitions in Section 7.2, we derive useful lemmas in Section 7.3 and show that a flow decomposition similar to the one for the traditional generalized flow problem is possible in the case of *convex* generalized flows as well. In Section 7.4, we consider the complexity and approximability of the convex generalized flow problem and show that the problem is both $\mathcal{NP}$-hard to solve and approximate. Afterwards, in Section 7.5, we present algorithms for the problem on graph classes with decreasing complexity. We present an algorithm that computes a maximum convex generalized preflow on general graphs in $\mathcal{O}(3^m \cdot m)$ time. Although such a maximum convex generalized preflow cannot be turned into a feasible flow on general graphs without further assumptions, it will be possible to derive a maximum convex generalized flow within the same time bound when restricting to acyclic graphs. Moreover, we show that we can improve this running time to $\mathcal{O}(2.707^m \cdot (m + n^2))$ in the case of series-parallel graphs and to $\mathcal{O}(2.404^m \cdot (m + n^2))$ time on extension-parallel graphs. Furthermore, we identify a special case of extension-parallel graphs for which the problem becomes solvable in polynomial time. Finally, in Section 7.6, we consider a variant of the problem in which the flows are restricted to be integral and present a pseudo-polynomial time algorithm for the problem on series-parallel graphs. An overview of the results of this chapter is given in Table 7.1 on page 191.

7.2 Preliminaries

We start by defining the convex generalized maximum flow problem in a directed graph $G = (V, E)$ with positive *edge capacities* $u_e > 0$ and *outflow functions* $g_e : [0, u_e] \to \mathbb{R}_{\geqslant 0}$ on the edges $e \in E$, and distinguished *source* $s \in V$ and *sink* $t \in V$. We assume that the outflow functions g_e fulfill the following property:

Assumption 7.1: The outflow functions g_e are strictly increasing continuous convex functions fulfilling $g_e(0) = 0$ for all $e \in E$. ◁

Note that, by standard results from analysis, Assumption 7.1 implies that the inverse functions g_e^{-1} are well-defined and continuous as well (cf., e.g., (Rudin, 1964)).

Definition 7.2 (Inflow, outflow, excess):
For any function $x : E \to \mathbb{R}_{\geqslant 0}$, the *inflow of an edge* $e \in E$ is given by $x_e := x(e)$ and the *outflow of edge* e is given by $g_e(x_e)$. Similarly, the *inflow (outflow) of a path* P equals the inflow (outflow) of the first (last) edge on P. For a node $v \in V$, the *inflow of* v is defined as $\sum_{e \in \delta^-(v)} g_e(x_e)$ and the *outflow of* v is given by $\sum_{e \in \delta^+(v)} x_e$. The *excess* of a node $v \in V$ with respect to x is given as $\text{excess}_x(v) := \sum_{e \in \delta^-(v)} g_e(x_e) - \sum_{e \in \delta^+(v)} x_e$. ◁

As in a traditional generalized flow, the outflow of an edge may differ from its inflow. Whereas the ratio between the outflow and the inflow of an edge is constant in traditional generalized flows, this ratio is now a non-decreasing function of the inflow.

Definition 7.3 (Pseudoflow, preflow, flow, flow value, maximum flow):
A function $x : E \to \mathbb{R}_{\geqslant 0}$ is called a (feasible) *convex generalized pseudoflow* (or just *pseudoflow*) if $x_e \leqslant u_e$ for all $e \in E$. If, in addition, $\text{excess}_x(v) \geqslant 0$ for all $v \in V \setminus \{s, t\}$, it is called a (feasible) *preflow*. If $\text{excess}_x(v) = 0$ for all $v \in V \setminus \{s, t\}$, the function is called

a (feasible) *convex generalized flow* (or just *flow*). The *flow value* of a (pre-)flow x is given by $val(x) := excess_x(t)$. A (pre-)flow of maximum flow value is called a *maximum (pre-)flow*. ◁

Note that a preflow is basically a flow that may send "too much" flow to some nodes such that flow conservation is not fulfilled. In some situations, such a relaxed flow may be much easier to compute than a feasible flow. In fact, while any preflow can be turned into a flow within polynomial time on acyclic graphs, such a transformation is uncomputable on general graphs as it will be shown in Section 7.5.1 and Section 7.5.2.

Using the above definitions, the convex generalized maximum flow problem is defined as follows:

Definition 7.4 (Convex Generalized Maximum Flow Problem (CGMFP)):

INSTANCE: Directed graph $G = (V, E)$ with source $s \in V$, sink $t \in V$, and non-negative capacities u_e and outflow functions g_e on the edges $e \in E$.

TASK: Determine a maximum flow.
◁

In addition to Assumption 7.1, we make the following assumptions on the structure of the underlying graph:

Assumption 7.5: For every node $v \in V \setminus \{s, t\}$, it holds that $\delta^+(v) \neq \emptyset$ and $\delta^-(v) \neq \emptyset$.
◁

Assumption 7.6: For every node $v \in V \setminus \{s, t\}$, there is at least one directed path from s to v or from v to t.
◁

Assumption 7.5 does not impose any restriction on the underlying model since the inflow and outflow of every node $v \in V \setminus \{s, t\}$ with $\delta^+(v) = \emptyset$ or $\delta^-(v) = \emptyset$ must equal zero due to flow conservation at v, which implies that the incident edges can be deleted in a preprocessing step. Similarly, Assumption 7.6 yields no restriction since the corresponding connected components that do not contain the sink t do not contribute to the flow value and can be deleted as well. Note that, for any instance of CGMFP, both assumptions can be established in $\mathcal{O}(n + m)$ time by performing a depth-first search and repeatedly deleting single nodes and edges. The resulting graph is connected, such that we can assume that $n \in \mathcal{O}(m)$ in the following.

An important special case of CGMFP considered throughout this chapter is the case of *quadratic outflow functions* of the form $g_e(x_e) = \alpha_e \cdot x_e^2$ for some positive constants $\alpha_e > 0$.

The problem CGMFP can be formulated as the following nonlinear program:

$$\max \sum_{e \in \delta^-(t)} g_e(x_e) - \sum_{e \in \delta^+(t)} x_e$$

$$\text{s.t.} \sum_{e \in \delta^-(v)} g_e(x_e) - \sum_{e \in \delta^+(v)} x_e = 0, \qquad \text{for all } v \in V \setminus \{s, t\},$$

$$0 \leqslant x_e \leqslant u_e, \qquad \text{for all } e \in E.$$

Note that this model is not solvable (in polynomial time) by standard solution methods for convex programs since a convex function is maximized (instead of minimized) over a set defined by convex and non-affine equality constraints (so the set of feasible solutions is non-convex in general). In the rest of this chapter, we will concentrate on combinatorial algorithms for the problem.

Moreover, note that, since we allow arbitrary strictly increasing continuous convex outflow functions g_e, it is not canonically clear how the functions are given in the input. In addition, neither the inflow nor the outflow of an edge in a maximum convex generalized flow can be assumed to be rational even if all capacities are integral: Consider, e.g., the instance that is depicted in Figure 7.1, in which the unique maximum generalized flow leads to an irrational inflow of $\sqrt[4]{2}$ and outflow of $\sqrt{2}$ on the first edge. Therefore, similar to the computational model used by Végh (2012) for the case

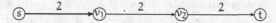

Figure 7.1: An example of a network with integral capacities where the first edge has an irrational inflow and outflow in the unique maximum flow. The labels on the edges represent the capacities. The outflow function of each edge is set to $g_e(x_e) = x_e^2$.

of concave generalized flows, we assume oracle access to the functions g_e and their inverses g_e^{-1} and the running time estimations for our algorithms provide bounds on the number of elementary arithmetic operations and oracle calls. We assume throughout this chapter that we can perform oracle calls returning the value $g_e(x_e)$ for some $x_e \in [0, u_e]$ and the value $g_e^{-1}(y_e)$ for some $y_e \in [0, g_e(u_e)]$ as well as elementary arithmetic operations on real numbers within infinite precision in constant time $\mathcal{O}(1)$. This is similar to standard assumptions in convex and semi-definite programming (cf. (Nesterov and Nemirovskii, 1994; Blum, 1998; Blum et al., 1989; Grötschel et al., 1993)).

7.3   Structural Results

In this section, we give insights into structural properties of convex generalized flows. On the one hand, we show that a flow decomposition theorem similar to the well-known flow decomposition theorem for traditional generalized flows is possible in the case of CGMFP as well. On the other hand, we derive a set of useful lemmas that will be used throughout the rest of this chapter.

The following result allows us to handle paths in the underlying network similarly as single edges:

Lemma 7.7:
Let $P = (e_1, \ldots, e_k)$ with $e_i = (v_i, v_{i+1})$, $i \in \{1, \ldots, k\}$, be a simple path and x a convex generalized pseudoflow in an acyclic graph $G = (V, E)$. When restricting the flow on each edge $e \notin P$ and the excess at each node $v \notin \{v_1, v_{k+1}\}$ to remain unchanged, the outflow of e_k can be described by a function $\bar{g}$ of the inflow of e_1. This function $\bar{g}$ is continuous, convex, and strictly increasing on the *set of feasible inflows* $[L_x(P), U_x(P)]$ *of* P, i.e., the set of all inflows $y_1 \geqslant 0$ of e_1 such that there exist values $y_2, \ldots, y_k \geqslant 0$ for which the function x' with $x'_{e_i} := y_i$ for $i \in \{1, \ldots, k\}$ and $x'_e := x_e$ for $e \notin P$ is a feasible pseudoflow with $\text{excess}_{x'}(v_i) = \text{excess}_x(v_i)$ for all $i \in \{2, \ldots, k\}$.

Proof: Consider the first two edges $e_1 = (v_1, v_2) \in P$ and $e_2 = (v_2, v_3) \in P$. Let δ be the excess that needs to be generated by the flows on e_1 and e_2 in order to maintain a total excess of $\text{excess}_x(v_2)$ at v_2:

$$
\delta := \text{excess}_x(v_2) - \left(\sum_{e \in \delta^-(v_2) \setminus \{e_1\}} g_e(x_e) - \sum_{e \in \delta^+(v_2) \setminus \{e_2\}} x_e \right)
$$

$$
= g_{e_1}(x_{e_1}) - x_{e_2}. \tag{7.1}
$$

Furthermore, let $[L_{e_i}, U_{e_i}] := [0, u_{e_i}]$ be the interval of feasible inflows of each edge e_i for $i \in \{1, \ldots, k\}$. When requiring the excess δ to remain constant, we can describe the outflow of e_2 depending on the inflow of e_1 by a strictly convex and increasing function $\bar{g}$ as well as the set of feasible inflows $[\bar{L}, \bar{U}]$ of the path (e_1, e_2) as follows:

According to equation (7.1), we get that it must hold that $x_{e_2} = g_{e_1}(x_{e_1}) - \delta$. As defined above, the inflow x_{e_2} of edge e_2 must both fulfill $x_{e_2} \geqslant L_{e_2}$ and $x_{e_2} \leqslant U_{e_2}$. Consequently, this is equivalent to the requirement that $x_{e_1} \geqslant g_{e_1}^{-1}(L_{e_2} + \delta)$ and that $x_{e_1} \leqslant g_{e_1}^{-1}(U_{e_2} + \delta)$. Since, furthermore, it must hold that $x_{e_1} \in [L_{e_1}, U_{e_1}]$, we get that $\bar{L} = \max\{L_{e_1}, g_{e_1}^{-1}(L_{e_2} + \delta)\}$ and $\bar{U} = \min\{U_{e_1}, g_{e_1}^{-1}(U_{e_2} + \delta)\}$. For each valid inflow $x \in [\bar{L}, \bar{U}]$, we can then express the outflow of g_{e_2} by $\bar{g}(x) = g_{e_2}(g_{e_1}(x) - \delta)$. Obviously,

$[\overline{L}, \overline{U}] \subseteq [L_{e_1}, U_{e_1}]$ and $[g_{e_1}(\overline{L}) - \delta, g_{e_1}(\overline{U}) - \delta] \subseteq [L_{e_2}, U_{e_2}]$. Thus, both g_{e_1} and g_{e_2} behave strictly convex and increasing for inflows in $[\overline{L}, \overline{U}]$. So, $\overline{g}$ is strictly convex and increasing as well.

By the above procedure, we are able to virtually join the first two edges of the path and to describe the outflow of the new edge by a strictly convex and increasing function. By induction, the claim follows. □

Using the results of Lemma 7.7, we are able to differentiate between full and empty paths – analogously to full and empty single edges:

Definition 7.8 (Full path, empty path):
Let $P = (e_1, \ldots, e_k)$ be a simple path and x a convex generalized pseudoflow in $G = (V, E)$. The path P is called *full (with respect to x)* if $x_{e_1} = U_x(P)$ and *empty (with respect to x)* if $x_{e_1} = L_x(P)$. ◁

Note that a path P is full if and only if there is at least one edge $e \in P$ with $x_e = u_e$. Analogously, we have that $x_e = 0$ for at least one edge $e \in P$ if and only if P is an empty path.

Example 7.9:
Consider the instance of CGMFP that is depicted in Figure 7.2, in which the outflow function and the capacity of each edge $e \in E$ is assumed to be $g_e(x_e) := x_e^2$ and $u_e := 15$, respectively. According to equation (7.1) in the proof of Lemma 7.7, we get that $\delta = 0 - (4 - 5) = 9 - 8 = 1$. It then follows that we can express the outflow of edge e_3 depending on the inflow of edge e_1 by the function $\overline{g}(x) = (x^2 - 1)^2 = x^4 - 2x^2 + 1$, which is convex, continuous, and increasing on the interval $[\overline{L}, \overline{U}]$ with $\overline{L} := \max\{0, \sqrt{0+1}\} = 1$ and $\overline{U} := \min\{15, \sqrt{15+1}\} = 4$. Note that for $x_{e_1} := \overline{L} = 1$ the path P is empty and the inflow of edge e_3 is zero. Conversely, for $x_{e_1} := \overline{U} = 4$ the inflow of edge e_3 equals 15 such that the path P is full. ◁

As it turns out, there is a flow decomposition that is similar to the one for traditional generalized flows, which will be shown in the following. To do so, we adapt the corresponding proof for traditional generalized flows with linear outflow functions from Goldberg et al. (1991).

One essential ingredient for the following results is the definition of a *subtraction of flow*. In the case of linear outflow functions (or traditional network flows), it is possible to subtract flow on some path P in the given graph (i.e., to reduce the flow on the edges of P) without influencing the flow on other paths $P' \neq P$. For general convex outflow functions, however, the subtraction becomes more complicated since the removal of flow on some edge e on a path P may reduce the flow value on other

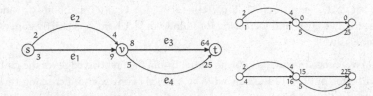

Figure 7.2: An example application of Lemma 7.7 on an instance of CGMFP with outflow functions $g_e := x_e^2$ and capacities $u_e := 15$ for each $e \in E$ (left). The situation for $x_{e_1} = \underline{L} = 1$ ($x_{e_1} = \overline{U} = 4$) is depicted on the upper (lower) right figure. The label on the tail (head) of each edge corresponds to the inflow (outflow) of the corresponding edge.

paths P' with $e \in P'$ as well. Nevertheless, we show that a subtraction similar to the traditional one is possible in our setting as well.

Definition 7.10 (Feasible subtraction):

Let x be a convex generalized pseudoflow in some graph $G = (V, E)$ and let $P = (e_1, \ldots, e_l)$ with $e_i = (v_i, v_{i+1})$, $i \in \{1, \ldots, l\}$, denote a (not necessarily simple) path in G. A function $\overline{x}: E \to \mathbb{R}_{\geqslant 0}$ with $\overline{x}_{e_i} > 0$ for each $i \in \{1, \ldots, l\}$ and $\overline{x}_e = 0$ for $e \notin P$ is called a *feasible subtraction on* P if $x_{e_i} - \overline{x}_{e_i} \geqslant 0$ for each $i \in \{1, \ldots, l\}$ and $\text{excess}_{x-\overline{x}}(v_i) = \text{excess}_x(v_i)$ for each $i \in \{2, \ldots, l\}$. ◁

Note that, while the excess at the nodes $v_2, \ldots, v_l$ is required to remain unchanged when reducing the flow on P by a feasible subtraction, the excess at the starting node v_1 and the end node v_{l+1} of the path may change during this procedure in case that $v_1, v_{l+1} \notin \{v_2, \ldots, v_l\}$.

Now consider a pseudoflow x and a simple path $P = (e_1, \ldots, e_l)$ with $e_i = (v_i, v_{i+1})$ and $x_{e_i} > 0$ for each $i \in \{1, \ldots, l\}$. According to Lemma 7.7, there is a flow x' with $x'_e = x_e$ for each $e \notin P$ and with $x'_{e_1} = L_x(P)$ as well as $\text{excess}_{x'}(v_i) = \text{excess}_x(v_i)$ for each $i \in \{2, \ldots, l\}$ such that the path P is empty. Clearly, the function $\overline{x} := x - x'$ is then a feasible subtraction on P fulfilling $\overline{x}_e = 0$ for $e \notin P$ and $\overline{x}_{e_i} > 0$ for each $i \in \{1, \ldots, l\}$. We call this subtraction $\overline{x}$ the *largest subtraction* on P in the following. Moreover, for a cycle C, we can consider C as a simple path with starting node and end node v for some node $v \in C$ and apply the above definition of a largest subtraction. However, the largest subtraction on C as well as the excess that is generated at the starting node v when reducing the flow on P by the largest subtraction may depend on the choice of the starting node.

Similar to traditional flows and generalized flows with linear outflow functions, a convex generalized pseudoflow x does not only decompose into s-t-paths, but also into cycles. As in the case of generalized flows with linear outflow functions, we distinguish three classes of cycles: flow generating cycles, flow absorbing cycles, and flow conserving cycles. The intuition is that a cycle is flow generating (flow absorbing) if removing the flow on the cycle yields a negative (positive) excess at some node on the cycle. If no excess is generated, the cycle is flow conserving.

Definition 7.11 (Flow generating cycle, flow absorbing cycle, flow conserving cycle):
Let C be a simple cycle in G, $v \in C$ a node on C, and x a convex generalized pseudoflow in G. The pair (C, v) is called a *flow generating cycle* (*flow absorbing cycle*) with respect to x if $\text{excess}_{x-\bar{x}}(v) < \text{excess}_x(v)$ ($\text{excess}_{x-\bar{x}}(v) > \text{excess}_x(v)$), where $\bar{x}$ denotes the largest subtraction on C when C is considered as a path with starting node (and end node) v. If $\text{excess}_{x-\bar{x}}(v) = \text{excess}_x(v)$, the pair (C, v) is called a *flow conserving cycle*. ◁

Note that, for generalized flows with linear outflow functions, Definition 7.11 can easily be seen to coincide with the standard definitions of flow generating and flow absorbing cycles independent of the choice of the starting node.

In the following, for some convex generalized pseudoflow x, let $\mathcal{D}(x)$ denote the set of nodes with *demand*, i.e., negative excess, and $\mathcal{S}(x)$ denote the set of nodes with *supply*, i.e., positive excess. Analogously to the elementary pseudoflows studied by Gondran and Minoux (1984) and Goldberg et al. (1991), we distinguish five types of *elementary subtractions*, where the type is determined by the graph induced by the set of edges on which the elementary subtraction is positive:

Definition 7.12 (Types of elementary subtractions):

Type I The largest subtraction on a simple path from a node in $\mathcal{D}(x)$ to a node in $\mathcal{S}(x)$.

Type II The largest subtraction on a simple path composed of a flow generating cycle (C, v) and a simple path from v to a node $w \in \mathcal{S}(x)$.

Type III The largest subtraction on a simple path composed of a flow absorbing cycle (C, v) and a simple path from a node $w \in \mathcal{D}(x)$ to v.

Type IV The largest subtraction on a flow conserving cycle (C, v).

Type V The largest subtraction on a simple path composed of a flow generating cycle (C_1, v_1) and a flow absorbing cycle (C_2, v_2) that are connected by a simple path from v_1 to v_2. ◁

Note that, by definition of feasible subtractions and flow conserving cycles, the excess may only change at the mentioned nodes in $\mathcal{D}(x)$ and $\mathcal{S}(x)$ when subtracting an elementary subtraction from the current pseudoflow x.

We are now ready to adopt the decomposition theorem for linear outflow functions from Goldberg et al. (1991) to the case of convex outflow functions.

Theorem 7.13 (Decomposition theorem for convex generalized pseudoflows):

A convex generalized pseudoflow x in a graph G can be decomposed into a sequence $(\overline{x}^{(1)}, \ldots, \overline{x}^{(k)})$ of $k \leqslant m$ elementary subtractions $\overline{x}^{(j)}$ such that $x_e = \sum_{j=1}^{k} \overline{x}_e^{(j)}$ for each $e \in E$.

Proof: We prove the claim by induction on the number p of edges with positive flow. If $p = 0$, then $x = 0$ and the claim trivially holds. Otherwise, let G' be the graph obtained from G by removing all edges with zero flow.

If G' is acyclic, we can find a simple path P from some node $v \in \mathcal{D}(x)$ to some node $w \in \mathcal{S}(x)$ with positive flow on each edge and the largest subtraction $\overline{x}$ on P is an elementary subtraction of Type I. Furthermore, as described above, $x - \overline{x}$ is a feasible convex generalized pseudoflow with at most $p - 1$ edges with positive flow. Thus, the claim follows by induction.

If G' is not acyclic, let $C = (v_1, \ldots, v_{l+1} = v_1)$ be a simple cycle in G'. We consider C as a simple path with starting node and end node v_1. As above, if we consider the largest subtraction $\overline{x}$ on C, then $x - \overline{x}$ is a feasible generalized pseudoflow that is zero on at least one edge in C. While the excess is maintained at each node v_i, $i \in \{2, \ldots, l\}$, the excess at v_1 may change by some amount $\delta \in \mathbb{R}$. If $\delta = 0$, the removal of flow on the cycle did not affect the remaining pseudoflow and $\overline{x}$ is an elementary subtraction of Type IV.

If $\delta < 0$, the pair (C, v_1) was a flow generating cycle. Since $x - \overline{x}$ is a valid generalized pseudoflow and has at most $p - 1$ edges with positive flow, we can apply the induction hypothesis and decompose $x - \overline{x}$. Since there was a demand at v_1, there are elementary subtractions of Type I and III in the decomposition of $x - \overline{x}$ that are responsible for the demand. These subtractions together with some appropriate fractions of $\overline{x}$ consequently correspond to elementary subtractions of Type II and Type V in G'. If $\delta > 0$, the pair (C, v_1) was a flow absorbing cycle and, by the same arguments as before, we obtain elementary subtractions of Type III and Type V.

In all three cases, the remaining pseudoflow is feasible again and contains at most $p - 1$ edges with positive flow. Hence, the claim follows by induction. $\qquad\square$

Note that, since the largest subtraction on a path or cycle depends on the current pseudoflow x, each elementary subtraction $\overline{x}^{(j)}$ will only be an elementary subtraction with respect to the pseudoflow obtained after all previous elementary subtractions $\overline{x}^{(1)}, \ldots, \overline{x}^{(j-1)}$ have already been subtracted from x. In particular, the order of the elementary subtractions within the sequence $(\overline{x}^{(1)}, \ldots, \overline{x}^{(k)})$ is important.

Also note that, in case that the pseudoflow x itself is a *flow*, the components in the decomposition for generalized pseudoflows with linear outflow functions obtained in Goldberg et al. (1991) are generalized flows again. Even if the pseudoflow x itself is a flow, however, each elementary subtraction in Theorem 7.13 will only be a feasible convex generalized flow with respect to different (strictly increasing, continuous, and convex) outflow functions given by $\overline{g}_e(x_e) := g_e(c_e + x_e) - g_e(c_e)$ with c_e denoting the remaining flow on edge e after the subtraction.

Finally, note that the computation of a flow decomposition using the recursive procedure presented in the proof of Theorem 7.13 is computationally intractable in general: For example, assume that a largest feasible subtraction $\overline{x}$ on a flow generating cycle (C, v_1) is being removed from the current pseudoflow x in some iteration of the procedure, which changes the excess at node v_1 by $\delta < 0$. A recursive application of the procedure to $x - \overline{x}$ yields elementary subtractions $\overline{x}^{(1)}, \ldots, \overline{x}^{(h)}$ of Type I or III that are responsible for the demand at v_1. However, for each such elementary subtraction $\overline{x}^{(j)}$ that generates an excess of $\delta^{(j)} < 0$ at v_1, we then need to determine the appropriate fraction of $\overline{x}$ that generates an excess of exactly $-\delta^{(j)}$ at v_1, i.e., we need to create a specific amount of flow on the cycle C. This is computationally intractable as we will see in the following section.

Example 7.14:

Figure 7.3 shows an exemplary flow and a possible decomposition into a sequence $(\overline{x}^{(1)}, \overline{x}^{(2)})$ of two elementary subtractions according to Theorem 7.13. In this example, all outflow functions are quadratic functions of the form $g_e(x_e) = \alpha_e \cdot x_e^2$ with positive constants $\alpha_e > 0$. Besides the source, which provides one unit of flow, the left-hand cycle is flow generating and creates two units of flow and the right-hand cycle is flow absorbing and consumes two units of flow (independently of the choice of the starting node). The rest of the flow is delivered to the sink.

In the example, there are several ways how to start the decomposition. Assume that we start by extracting an elementary subtraction $\overline{x}^{(1)}$ of Type I that is depicted on the lower left of Figure 7.3. While the inflow at node v_1 is reduced by one unit, the inflow at v_2 sinks by 9 units. The elementary subtraction $\overline{x}^{(1)}$ is a feasible convex generalized flow when changing the outflow function of the edge $e = (v_1, v_2)$ to $\overline{g}_e(x_e) = g_e(4 + x_e) - g_e(4) = (4 + x_e)^2 - 16 = x_e^2 + 8 \cdot x_e$. In this case, we are done since the remaining flow is an elementary subtraction $\overline{x}^{(2)}$ of Type V (shown on the right side of Figure 7.3). ◁

We end this section with a useful property of paths in acyclic graphs:

Lemma 7.15:

Let x be a convex generalized flow in an acyclic graph $G = (V, E)$ and let P_1, P_2 be two

Feasible flow x:

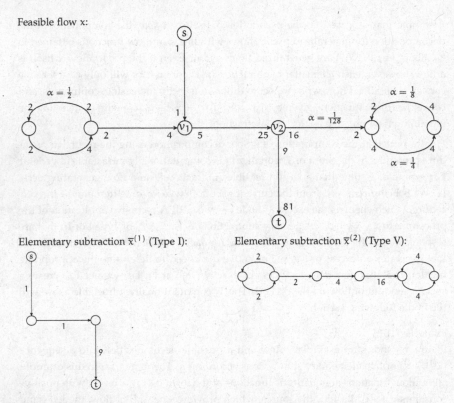

Figure 7.3: A sample flow and a possible decomposition. In the upper figure, the tail (head) of each edge is labeled with the corresponding inflow (outflow). The constant α_e in the outflow function $g_e(x_e) = \alpha_e \cdot x_e^2$ is equal to one if not given explicitly. In the lower figures, each edge is labeled with the amount of flow subtracted. The given flow can be decomposed into an elementary subtraction $\bar{x}^{(1)}$ of Type I (bottom left) and an elementary subtraction $\bar{x}^{(2)}$ of Type V (bottom right).

edge-disjoint v_1-v_2-paths for some nodes $v_1, v_2 \in V$. If both paths are neither full nor empty, then there exists a convex generalized flow x' with $val(x') = val(x)$ for which at least one of the paths P_1 and P_2 is full or empty.

Proof: By Lemma 7.7, we can describe the outflow of each P_i, $i \in \{1, 2\}$, by a convex function g_i of its inflow x_i. Writing $\varepsilon := x_1 + x_2$, the total amount of flow arriving at v_2 via P_1 and P_2 is then given as

$$f(x_1) := g_1(x_1) + g_2(\varepsilon - x_1)$$

for

$$x_1 \in [L, U] := [L_x(P_1), U_x(P_1)] \cap [\varepsilon - U_x(P_2), \varepsilon - L_x(P_2)].$$

Since both g_1 and g_2 are convex and increasing, so is f. Thus, the maximum of f on $[L, U]$ is obtained at either L or U and, hence, at a boundary of $[L_x(P_1), U_x(P_1)]$ or $[\varepsilon - U_x(P_2), \varepsilon - L_x(P_2)]$, which means that we can obtain an excess of at least $excess_x(v_2)$ at v_2 by choosing $x_1 = L$ or $x_1 = U$. By definition of L, U, and ε, this corresponds to turning P_1 or P_2 into a full or empty path.

In case that the excess of v_2 was increased by the above procedure, we can regain the old excess at v_2 by reducing x_1 or x_2 appropriately: If $L_{x'}(P_i) < x'_i < U_{x'}(P_i)$ for the new pseudoflow x' and some $i \in \{1,2\}$, we can reduce x'_i until either $x'_i = L_{x'}(P_i)$ or the excess at v_2 attains its old value. In the former case, the path P_i becomes empty and we can reduce the flow on the other path until the excess at v_2 attains its old value. If one path P_i is full and the other path is empty, we can simply reduce x'_i until the excess at v_2 attains its old value.

Note that this procedure for regaining the old excess at v_2 may create a positive excess at v_1. This excess can be eliminated by reducing the flow on some of the paths that transport flow from the source s to v_1 in a similar way as above. Hence, only the excess of the source s is changed in the overall procedure. In particular, we obtain a feasible flow with the same flow value as x, which proves the claim. $\qquad\square$

Example 7.16:
Figure 7.4 shows an application of Lemma 7.15 on a small acyclic graph. None of the two paths (edges) between v_1 and v_2 is full or empty, but the flow is optimal since the single edge leading to the sink t is filled to its capacity. According to the proof of Lemma 7.15, we can redistribute the flow on the lower path to the upper path, which creates an outflow of value 16 and, thus, an excess of 8 at v_2. We then reduce the inflow of the upper path to $\sqrt{8}$ in order to regain flow conservation at v_2. This, in turn, leads to an excess of $4 - \sqrt{8} > 0$ at v_1, which can be resolved by reducing the inflow of the edge (s, v_1) from 2 to $\sqrt[4]{8} < 2$. $\qquad\triangleleft$

7.4 Complexity and Approximability

In this section, we consider the complexity and approximability of the convex generalized flow problem. As it turns out – in contrast to traditional generalized flows and generalized flows with concave outflow functions – the problem is $\mathcal{NP}$-hard to solve and approximate in general, which will be shown in Section 7.4.1 and 7.4.2, respectively.

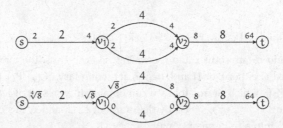

Figure 7.4: A sample application of Lemma 7.15. The number above each edge denotes the capacity of the corresponding edge. The outflow function of every edge e is $g_e(x_e) = x_e^2$. The upper scenario shows an optimal solution in which none of the two paths (edges) from v_1 to v_2 is full or empty. By applying the steps from the proof of Lemma 7.15, we obtain another optimal solution in which the lower path between v_1 and v_2 is empty.

As noted above, we assume the underlying computational model to cohere with the Blum-Shub-Smale model in which we can perform arithmetic operations on irrational numbers within infinite precision in constant time. For this model, an independent theory of $\mathcal{NP}$-completeness has been developed and the connection between this theory and the traditional theory based on the RAM model is unclear. For example, the well-known *traveling salesman problem*, which is $\mathcal{NP}$-complete in the standard RAM model, is not known to be $\mathcal{NP}$-complete in the BSS model. Nevertheless, we want to stress that we use the BSS model only for the sake of simplicity. It can be easily seen that the upcoming algorithms and complexity results are also valid when restricting our considerations to outflow functions that map rational numbers to rational numbers. Hence, the complexity results presented in this section will be based on the traditional theory of $\mathcal{NP}$-completeness in the RAM model. We refer the reader to (Blum, 1998) for further details on the BSS model and the connection to the RAM model.

7.4.1 Complexity

We start by proving that the convex generalized maximum flow problem is strongly $\mathcal{NP}$-hard to solve on general graphs.

Theorem 7.17:
CGMFP is strongly $\mathcal{NP}$-hard to solve, even if all outflow functions are quadratic outflow functions of the form $g_e(x_e) = x_e^2$, the capacities are integral, and the graph is bipartite and acyclic.

Proof: We use a reduction from the EXACTCOVERBY3SETS problem, which is known to be strongly $\mathcal{NP}$-complete (cf. (Garey and Johnson, 1979, Problem SP2)):

INSTANCE: Set X with 3q elements and a collection $\mathcal{C} = \{C_1, \ldots, C_k\}$ of 3-element subsets of X.

QUESTION: Does there exist a subcollection $\mathcal{C}' \subseteq \mathcal{C}$ such that every element $j \in X$ is contained in exactly one of the subsets in $\mathcal{C}'$?

Given an instance of EXACTCOVERBY3SETS, we construct a network for CGMFP as follows:

We introduce a single source s and sink t as well as a node s', which is reachable from s via a single edge with capacity 3q. For each subset $C_i \in \mathcal{C}$, $i \in \{1, \ldots, k\}$, we insert a node v_i and an edge between s' and v_i with capacity $3^2 q$. Furthermore, we introduce a node v'_j for each $j \in X$, which is reachable from every v_i with $j \in C_i$ via an edge with capacity $3^3 q^2$. Finally, we connect each v'_j to the sink t by an edge with capacity $3^6 q^4$. All outflow functions are set to $g_e(x_e) := x_e^2$. The resulting network for the set $X = \{1, \ldots, 9\}$ and the collection $\mathcal{C} = \{\{1,2,4\}, \{2,3,4\}, \{3,5,8\}, \{4,6,7\}, \{6,7,9\}\}$ is shown in Figure 7.5.

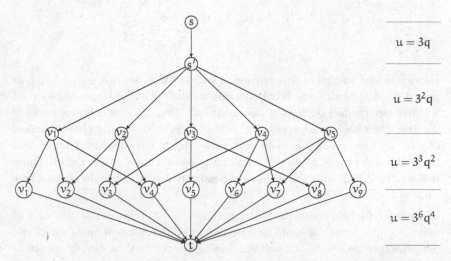

$$u = 3q$$

$$u = 3^2 q$$

$$u = 3^3 q^2$$

$$u = 3^6 q^4$$

Figure 7.5: The resulting network for a given instance of EXACTCOVERBY3SETS with $q = 3$, $X = \{1, \ldots, 9\}$ and $\mathcal{C} = \{\{1,2,4\}, \{2,3,4\}, \{3,5,8\}, \{4,6,7\}, \{6,7,9\}\}$. On the right hand side, the capacities of the edges in each level of the graph are depicted. The outflow function of every edge is given by $g_e(x_e) := x_e^2$.

We now show that there exists a convex generalized flow of value at least $3^{13}q^9$ if and only if the given instance of ExactCoverBy3Sets is a Yes-instance.

Suppose that there is a convex generalized flow x in the constructed network with flow value val$(x) \geqslant 3^{13}q^9$. The maximum amount of flow that may arrive at s' is 3^2q^2 since the capacity of the edge between s and s' is $3q$ and the outflow function is given as $g(x_e) = x_e^2$. Furthermore, we claim that the total amount of flow arriving at the nodes v_i, $i \in \{1, \ldots, k\}$, is at most 3^4q^3 and this value is achieved if and only if the inflow of 3^2q^2 at s' is distributed equally on exactly q of the edges (s', v_i). To see this, consider a set $I \subseteq \{1, \ldots, k\}$ with $|I| = q$. By sending $x_i := 3^2q$ units of flow to v_i, $i \in I$, and $x_i := 0$ units to the remaining v_i, $i \in \{1, \ldots, k\} \setminus I$, flow conservation at node s' is fulfilled and, in total, $\sum_{i \in I} g(x_i) = q \cdot (3^2q)^2 = 3^4q^3$ units of flow reach the nodes v_i. Conversely, consider a flow x' that uses more than q of the edges (s', v_i). Then, there are at least two nodes v_{i_1} and v_{i_2} such that $x'_{i_1}, x'_{i_2} \in (0, 3^2q)$. Without loss of generality, we can assume that $x'_{i_1} \geqslant x'_{i_2}$. This flow cannot yield the highest possible amount of flow arriving at the nodes v_i since increasing x'_{i_1} and decreasing x'_{i_2} by some positive amount $\varepsilon \leqslant \min\{3^2q - x'_{i_1}, x'_{i_2}\}$ leads to a strictly higher amount of flow arriving at the nodes v_{i_1}, v_{i_2}:

$$(x'_{i_1} + \varepsilon)^2 + (x'_{i_2} - \varepsilon)^2 = (x'_{i_1})^2 + (x'_{i_2})^2 + \underbrace{2\varepsilon(x'_{i_1} - x'_{i_2})}_{\geqslant 0} + \underbrace{2\varepsilon^2}_{>0}$$

$$> (x'_{i_1})^2 + (x'_{i_2})^2.$$

Hence, the total amount of flow arriving at each of the nodes v_i, $i \in \{1, \ldots, k\}$, is at most 3^4q^3. Additionally, this is only the case if exactly q out of k nodes v_i are used which in turn produce outflows of value 3^4q^2 each. Since the sum of the capacities of the three edges leaving each v_i is $3 \cdot 3^3q^2 = 3^4q^2$, every such edge must have an inflow of value 3^3q^2 and produce an outflow of value 3^6q^4 in this situation. On the other hand, note that each of the $3q$ edges leading to the sink must receive the maximum inflow of value 3^6q^4 since this is the only possibility how to obtain the claimed flow value of $3^{13}q^9 = 3q \cdot (3^6q^4)^2$.

In summary, the total flow arriving at the nodes v'_j, $j \in X$, is at most $3q \cdot 3^6q^4 = 3^7q^5$ which, in turn, is the minimum flow needed to achieve the flow value val(x) at t. Since this flow can only be achieved by selecting q out of the k nodes v_i whose sets of adjacent nodes v'_j are pairwise disjoint and cover all nodes v'_j, we obtain a solution to ExactCoverBy3Sets by identifying the used nodes v_i with the given sets $C_i \in \mathcal{C}$.

Conversely assume that there exists a solution $\mathcal{C}' \subseteq \mathcal{C}$ for the given instance of ExactCoverBy3Sets. By sending $3q$ units of flow to s' and 3^2q units of flow to each of the nodes v_i corresponding to the subsets $C_i \in \mathcal{C}'$, we achieve q inflows of value $(3^2q)^2 = 3^4q^2$ each at the nodes v_i. The flow can further be distributed to the nodes v'_j, $j \in X$, in packages of $3^3 q^2$ each. Thus, we get an inflow of value $(3^3 q^2)^2 = 3^6 q^4$ at every node v'_j. Since each element $j \in X$ is contained in exactly one of the sets $C_i \in \mathcal{C}'$, each of these packages can further be sent to t producing outflows of $(3^6 q^4)^2 = 3^{12} q^8$ each. Consequently, since there are 3q edges leading to the sink, we achieve a flow value of $3q \cdot 3^{12} q^8 = 3^{13} q^9$. □

Theorem 7.17 shows that, unless $\mathcal{P} = \mathcal{NP}$, one cannot expect to find an algorithm that solves the problem exactly and runs in polynomial time. The reason why the theorem only claims $\mathcal{NP}$-hardness instead of $\mathcal{NP}$-completeness (in the standard Turing machine model) is that the problem CGMFP need not always have rational solutions (of polynomial size). In the Blum-Shub-Smale model (Blum et al., 1989), however, CGMFP is readily seen to be in $\mathcal{NP}$ since the feasibility and the flow value of a given convex generalized flow can then be checked easily in (oracle) polynomial time. Nevertheless, we want to stress that, in the presented reduction, a YES-instance of CGMFP contains only integral numbers and can, thus, be verified in polynomial time using the standard Turing machine model.

Theorem 7.18:
CGMFP on extension-parallel graphs is weakly $\mathcal{NP}$-hard to solve, even if all outflow functions are quadratic functions of the form $g_e(x_e) = \alpha_e \cdot x_e^2$ with integral constants $\alpha_e > 0$ and all capacities are integral.

Proof: We use a reduction from the weakly $\mathcal{NP}$-complete SUBSETSUM problem, which is defined as follows (cf. (Garey and Johnson, 1979, Problem SP13)):

INSTANCE: Finite set $A = \{a_1, \ldots, a_k\}$ of k positive integers and a positive integer B.

QUESTION: Is there a subset $I \subseteq \{1, \ldots, k\}$ such that $\sum_{i \in I} a_i = B$?

Given an instance of SUBSETSUM, we construct a network for CGMFP by introducing three nodes s, v, and t. Between s and v, we insert an edge e_0 with capacity 1 and outflow function $g_{e_0}(x_{e_0}) := B \cdot x_{e_0}^2$. Additionally, we introduce an edge e_i between v and t with capacity a_i and outflow function $g_{e_i}(x_{e_i}) := \frac{\pi}{a_i} \cdot x_{e_i}^2$ for each $i \in \{1, \ldots, k\}$, where $\pi := \prod_{j=1}^{k} a_j$. Note that the factors $\alpha_{e_i} := \frac{\pi}{a_i} = \prod_{j \neq i} a_j$ are integral and the resulting graph is extension-parallel. The constructed network is shown in Figure 7.6.

We now show that there exists a convex generalized flow of value at least $B \cdot \pi$ if and only if the given instance of SUBSETSUM is a YES-instance.

Suppose that there is a convex generalized flow x of value $val(x) \geq B \cdot \pi$ in the constructed network. Note that, for each edge e_i between v and t with capacity a_i and inflow x_{e_i}, the outflow is given by $\frac{\pi}{a_i} \cdot x_{e_i}^2$, which evaluates to $\pi \cdot x_{e_i}$ if $x_{e_i} = a_i$ and to some smaller multiple of x_{e_i} if $x_{e_i} < a_i$. Hence, since the maximum possible outflow

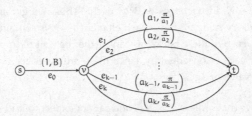

Figure 7.6: The constructed network for the given instance of SUBSETSUM. The label on each edge e denotes the capacity u_e and the factor α_e, respectively.

of e_0 is B, the flow value of $val(x) \geqslant B \cdot \pi$ implies that B units of flow must arrive at v and each edge among $e_1, \ldots, e_k$ that has positive inflow satisfies $x_{e_i} = a_i$. Hence, the B units of flow arriving at v are distributed to some edges e_i, $i \in I$, for some subset $I \subseteq \{1, \ldots, k\}$ that satisfies $\sum_{i \in I} a_i = B$, i.e, the given instance of SUBSETSUM is a YES-instance.

Conversely assume that there exists a solution $I \subseteq \{1, \ldots, k\}$ of the given instance of SUBSETSUM, i.e., $\sum_{i \in I} a_i = B$. By sending $x_{e_0} := 1$ units of flow along e_0 (which amounts to an inflow of value B at v), $x_{e_i} := a_i$ units of flow over the edges e_i for $i \in I$, and $x_{e_j} := 0$ units along the edges e_j for $j \in \{1, \ldots, k\} \setminus I$, we obtain flow conservation at v and get a feasible convex generalized flow x of value

$$val(x) = \sum_{i \in \{1,\ldots,k\}} g_{e_i}(x_{e_i}) = \sum_{i \in I} g_{e_i}(a_i) = \sum_{i \in I} \frac{\pi}{a_i} \cdot a_i^2 = \sum_{i \in I} \pi \cdot a_i$$
$$= B \cdot \pi. \qquad \square$$

7.4.2 Approximability

Since the convex generalized flow problem is strongly $\mathcal{NP}$-hard to solve on general graphs, as it was shown in the preceding section, one might still hope for efficient approximation algorithms for the problem. However, as it turns out, the problem is $\mathcal{NP}$-hard to approximate as well, even on simple graph classes:

Theorem 7.19:
CGMFP is $\mathcal{NP}$-hard to approximate within constant factors, even on extension-parallel graphs.

Proof: For $\varepsilon \in (0, 1)$, suppose that there was an $(1 - \varepsilon)$-approximation algorithm for CGMFP that computes a feasible flow x with $(1 - \varepsilon) \cdot val(x^*) \leqslant val(x) \leqslant val(x^*)$ in polynomial time, where x^* is a maximum convex generalized flow. We show that this $(1 - \varepsilon)$-approximation algorithm allows us to decide if any instance of SUBSETSUM is a YES-instance, which is a contradiction unless $\mathcal{P} = \mathcal{NP}$.

Let $(\{a_1, \ldots, a_k\}, B)$ denote an instance of SUBSETSUM. Without loss of generality, we may assume that $a_i \geqslant 2$ for each $i \in \{1, \ldots, k\}$ since we can otherwise multiply each of the values a_i and B by two. Similar to the proof of Theorem 7.18, we construct a network for CGMFP by introducing four nodes s, v, w, and t. Between s and v, we insert an edge e_0 with capacity 1 and outflow function $g_{e_0}(x_{e_0}) := B \cdot x_{e_0}$. For each $i \in \{1, \ldots, k\}$, we introduce an edge e_i between v and w with capacity a_i and outflow function

$$g_{e_i}(x_{e_i}) := \begin{cases} \frac{x_{e_i}}{a_i - 1} \cdot \frac{1}{2}, & \text{if } x_{e_i} \leqslant a_i - 1, \\ (x_{e_i} - (a_i - 1)) \cdot (a_i - \frac{1}{2}) + \frac{1}{2}, & \text{else.} \end{cases}$$

Note that the functions g_{e_i} are continuous, increasing, and convex for each $i \in \{1, \ldots, k\}$: Let $g_{e_i}^{(1)}(x_{e_i}) := \frac{x_{e_i}}{a_i - 1} \cdot \frac{1}{2}$ and $g_{e_i}^{(2)}(x_{e_i}) := (x_{e_i} - (a_i - 1)) \cdot (a_i - \frac{1}{2}) + \frac{1}{2}$ denote the two linear segments of g_{e_i}. It holds that $g_{e_i}^{(1)}(a_i - 1) = \frac{1}{2} = g_{e_i}^{(2)}(a_i - 1)$, which shows continuity of g_{e_i}. Moreover, it holds that

$$0 < \frac{d}{dx_{e_i}} g_{e_i}^{(1)}(x_{e_i}) = \frac{1}{2(a_i - 1)} < 1 < a_i - \frac{1}{2} = \frac{d}{dx_{e_i}} g_{e_i}^{(2)}(x_{e_i}),$$

which shows both convexity and monotonicity of g_{e_i}.

Moreover, we insert an edge $\bar{e}$ between w and t with capacity B and outflow function

$$g_{\bar{e}}(x_{\bar{e}}) := \begin{cases} \frac{x_{\bar{e}}}{B - \frac{1}{2}} \cdot (1 - \varepsilon) \frac{B}{4}, & \text{if } x_{\bar{e}} \leqslant B - \frac{1}{2}, \\ (x_{\bar{e}} - (B - \frac{1}{2})) \cdot (1 + \varepsilon) \frac{B}{2} + (1 - \varepsilon) \frac{B}{4}, & \text{else.} \end{cases}$$

Similar to the functions g_{e_i}, the function $g_{\bar{e}}$ is continuous, increasing, and convex as well: For $g_{\bar{e}}^{(1)}(x_{\bar{e}}) := \frac{x_{\bar{e}}}{B - \frac{1}{2}} \cdot (1 - \varepsilon) \frac{B}{4}$ and $g_{\bar{e}}^{(2)}(x_{\bar{e}}) := (x_{\bar{e}} - (B - \frac{1}{2})) \cdot (1 + \varepsilon) \frac{B}{2} + (1 - \varepsilon) \frac{B}{4}$, it holds that $g_{\bar{e}}^{(1)}(B - \frac{1}{2}) = (1 - \varepsilon) \frac{B}{4} = g_{\bar{e}}^{(2)}(B - \frac{1}{2})$, which shows continuity of $g_{\bar{e}}$. Furthermore, since

$$0 < \frac{d}{dx_{\bar{e}}} g_{\bar{e}}^{(1)}(x_{\bar{e}}) = \frac{(1 - \varepsilon) \cdot \frac{B}{4}}{B - \frac{1}{2}} = \frac{(1 - \varepsilon)}{2B - 1} \cdot \frac{B}{2} < \frac{B}{2} < (1 + \varepsilon) \frac{B}{2} = \frac{d}{dx_{\bar{e}}} g_{\bar{e}}^{(2)}(x_{\bar{e}}),$$

the function $g_{\bar{e}}$ is increasing and convex as well.

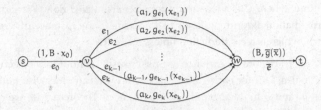

Figure 7.7: The constructed network for the given instance of SUBSETSUM. The label on each edge e denotes the capacity u_e and the outflow function g_e, respectively.

The resulting network is depicted in Figure 7.7. We now show that the flow value of a maximum convex generalized flow equals B if there is a solution to the given instance of SUBSETSUM and less than $(1 - \varepsilon) \cdot B$ else.

Let $I \subseteq \{1, \ldots, k\}$ be a solution to the given instance of SUBSETSUM, i.e., $\sum_{i \in I} a_i = B$. By sending $x_{e_0} := 1$ unit of flow through e_0, $x_{e_i} := a_i$ units of flow through edge e_i for $i \in I$, $x_{e_j} := 0$ units of flow through edge e_j, $j \in \{1, \ldots, k\} \setminus I$, and $x_{\bar{e}} := B$ units of flow through $\bar{e}$, we get a feasible convex generalized flow x with flow value $\text{val}(x) = B$, which is clearly maximum.

Now suppose that there is no solution to the given instance of SUBSETSUM and let $I := \{i \in \{1, \ldots, k\} : x^*_{e_i} > 0\}$, where x^* is a maximum convex generalized flow in the constructed instance of CGMFP. Clearly, if $\sum_{i \in I} a_i \leq B - 1$, the flow arriving at node w amounts to at most $\sum_{i \in I} g_{e_i}(a_i) = \sum_{i \in I} a_i \leq B - 1$, which causes a flow value of less than $(1 - \varepsilon) \cdot \frac{B}{4} < (1 - \varepsilon) \cdot B$ after passing edge $\bar{e}$ due to the definition of $g_{\bar{e}}$. If $\sum_{i \in I} a_i \geq B + 1$, there is exactly one $i \in I$ with $0 < x^*_{e_i} < a_i$ without loss of generality according to Lemma 7.15 since there are only at most B units of flow arriving at node v. Moreover, since all of the values a_i are integral, it holds that $x^*_{e_i} \leq a_i - 1$ such that $g_i(x^*_{e_i}) \leq \frac{1}{2}$. Then, however, the amount of flow that arrives at node w is given by $\sum_{j \in I} x^*_j \leq \sum_{j \in I \setminus \{i\}} a_j + \frac{1}{2} \leq B - \frac{1}{2}$, which implies that the maximum flow value $\text{val}(x^*)$ is less than $(1 - \varepsilon) \cdot \frac{B}{4} < (1 - \varepsilon) \cdot B$.

Hence, the flow value of a maximum convex generalized flow in the constructed network is B if there is a solution to the given instance of SUBSETSUM and less than $(1 - \varepsilon) \cdot B$ else. Thus, any $(1 - \varepsilon)$-approximation algorithm returns a solution x with flow value $\text{val}(x) \geq (1 - \varepsilon) \cdot B$ if and only if the underlying instance of SUBSETSUM is a YES-instance, which proves the theorem. □

7.5 Exact Algorithms

As it was shown in the preceding section, the problem CGMFP is both $\mathcal{NP}$-hard to solve and to approximate. Thus, unless $\mathcal{P} = \mathcal{NP}$, we will not be able to solve the problem in polynomial time. However, we are able to derive exponential-time exact algorithms that compute maximal generalized (pre-)flows. In this section, we consider different graph classes with decreasing structural complexity. We will be able to derive more efficient algorithms with decreasing complexity of the underlying graph. Moreover, in Section 7.5.5, we introduce a special case of extension-parallel graphs for which a maximum convex generalized flow can be computed in polynomial time.

7.5.1 General Graphs

In this section, we present an exponential-time algorithm that computes a maximum preflow on general graphs in $\mathcal{O}(3^m \cdot m)$ time. We start by proving several auxiliary results.

The proof of the following well-known fact is provided for the sake of completeness:

Lemma 7.20:
A full-dimensional polytope in $\mathbb{R}^n$ has at least $n + 1$ facets.

Proof: Let $\{x \in \mathbb{R}^n : Ax \leqslant b\}$ be an non-redundant formulation of a full-dimensional polytope $P \subseteq \mathbb{R}^n$, $A \in \mathbb{R}^{m \times n}$, $b \in \mathbb{R}^m$. Since P is full-dimensional and bounded, it contains at least two distinct extreme points $x^{(1)}$ and $x^{(2)}$. Each of these extreme points can be determined by setting a set of n inequalities to equality. Thus, since $x^{(1)} \neq x^{(2)}$, the formulation has at least $n + 1$ inequalities. It is well known that, in such a non-redundant formulation, there is a one-to-one correspondence between the inequalities and the facets of P, see e.g. (Schrijver, 1998). $\square$

Consider a partition (L, T, U) of the edge set E into three sets L, T, and U, where T forms a spanning tree of the graph G. Similar to the network simplex algorithm for the traditional minimum cost flow problem (cf. Section 4.3), we refer to this partition (L, T, U) as a *basis structure* in the following. We refer to any convex generalized preflow x fulfilling $x_e = 0$ for each $e \in L$ and $x_e = u_e$ for each $e \in U$ as a *preflow corresponding to the basis structure*. As in the case of traditional and budget-constrained minimum cost flows, we can restrict our considerations to such preflows:

Proposition 7.21:
Let x be a convex generalized preflow in a graph $G = (V, E)$. Then there exists

a convex generalized preflow x' corresponding to a basis structure (L, T, U) with $val(x') \geqslant val(x)$.

Proof: For the given convex generalized preflow x, consider the partition of the edge set given as $L := \{e \in E : x_e = 0\}$, $U := \{e \in E : x_e = u_e\}$, and $T := \{e \in E : x_e \in (0, u_e)\}$. If the subgraph that is induced by the edges in T does not contain any cycle, the claim clearly follows since we can add edges from L or U to T until the edges in T form a spanning tree of G.

Now let $C = (e_1, \ldots, e_k)$ be a (possibly undirected) cycle in G such that $e_i \in T$ and $x_{e_i} \in (0, u_{e_i})$ for each $i \in \{1, \ldots, k\}$. In the following, we refer to such a cycle as a T-*cycle*. We show that there also exists a preflow x' with $val(x') \geqslant val(x)$ in which the flow on C is rerouted in a way such that at least one edge on C belongs to L or U. By a repeated application of these arguments, the claim then follows.

Let $(P_1, \ldots, P_\kappa)$ denote the partition of C into maximal directed subpaths. We replace each of the subpaths P_i by a single edge $\bar{e}_i$. The outflow of each such edge $\bar{e}_i$ can then be described by a convex function $\bar{g}_i$ of its inflow according to Lemma 7.7. If $\kappa = 1$, i.e., C is a directed cycle, we can set the flow entering the cycle at the starting node v of P_1 to $U_x(P_1)$ or $L_x(P_1)$ depending on whether (C, v) is a flow generating cycle or a flow absorbing cycle for x. Since the excess then increases at v and remains constant at every other node, the claim follows.

Now let C be an undirected cycle. By construction, in the resulting (undirected) cycle $\bar{C} = (\bar{e}_1, \ldots, \bar{e}_\kappa)$, the directions of the edges alternate. The situation before and after this procedure is depicted in Figure 7.8.

Note that the number κ of nodes and edges on $\bar{C}$ is even since the direction of the edges changes at each node. Moreover, we may assume without loss of generality that $\kappa \geqslant 4$ since, for the case $\kappa = 2$, the cycle consists of two parallel paths and we can proceed as in the proof of Lemma 7.15 in order to make one of the paths full or empty while only generating positive excess at the end node. Furthermore, each node on $\bar{C}$ has either two incoming edges or two outgoing edges in $\bar{C}$ and we assume that the edges and nodes are labeled as in Figure 7.8, i.e., nodes with odd index have two outgoing edges and nodes with even index have two incoming edges. Furthermore, edges with odd index j start from $\bar{v}_j$ and head to $\bar{v}_{j+1}$, while edges with even index l head from $\bar{v}_{l+1}$ to $\bar{v}_l$. Let $\bar{g}_i$ be the function describing the outflow of $\bar{e}_i$ for $i \in \{1, \ldots, \kappa\}$, which is convex according to the above explanations. Furthermore, let s_j denote the sum of the inflows of the edges $\bar{e}_{j-1}$ and $\bar{e}_j$ in the given preflow x for $j \in \{1, 3, 5, \ldots, \kappa - 1\}$, i.e., the flow leaving node $\bar{v}_j$ along the edges of the cycle $\bar{C}$. Similarly, let d_l denote the flow arriving at $\bar{v}_l$ via $\bar{e}_{l-1}$ and $\bar{e}_l$ in x, $l \in \{2, 4, \ldots, \kappa\}$. Note that, due to notational convenience, we avoid using modulo-functions, i.e., whenever

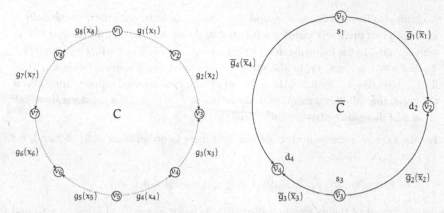

Figure 7.8: Replacing each maximal directed subpath of a cycle by a single edge yields a cycle in which the directions of the edges alternate. The left figure shows the situation before, the right figure after the procedure. The new outflow functions $\overline{g}_i$ are convex on the set of feasible inflows.

an index evaluates to 0 or $\kappa + 1$, it should be κ or 1, respectively. Moreover, we will always denote odd indices by j and even indices by l.

Our aim is to distribute the flows s_j leaving the odd nodes $\overline{v}_j$ onto $\overline{e}_{j-1}$ and $\overline{e}_j$ in a way such that the inflow of each even node $\overline{v}_l$ is at least d_l, while the flow on at least one edge $\overline{e}_i$ lies in $\{L_x(P_i), U_x(P_i)\}$ (which means that at least one edge $e \in P_i$ will have $x'_e = 0$ or $x'_e = u_e$ as remarked above after Definition 7.8). To do so, we formulate a system of nonlinear inequalities in the inflows y_j of the odd edges $\overline{e}_j$. The inflow y_l of each even edge $\overline{e}_l$ can then be expressed as $y_l = s_{l+1} - y_{l+1}$. We then have the boundary conditions $y_j \in [L_x(P_j), U_x(P_j)]$ and $s_j - y_j \in [L_x(P_{j-1}), U_x(P_{j-1})]$ or, equivalently,

$$y_j \in [L_j, U_j] := [L_x(P_j), U_x(P_j)] \cap [s_j - U_x(P_{j-1}), s_j - L_x(P_{j-1})].$$

Hence, we want to find a solution to the following system of nonlinear inequalities for which $y_j \in \{L_j, U_j\}$ for at least one $j \in \{1, 3, \ldots, \kappa - 1\}$:

$$\overline{g}_1(y_1) + \overline{g}_2(s_3 - y_3) \geqslant d_2,$$
$$\overline{g}_3(y_3) + \overline{g}_4(s_5 - y_5) \geqslant d_4,$$
$$\vdots$$
$$\overline{g}_{\kappa-1}(y_{\kappa-1}) + \overline{g}_\kappa(s_1 - y_1) \geqslant d_\kappa,$$
$$y_j \in [L_j, U_j], j \in \{1, 3, \ldots, \kappa - 1\}.$$

According to the definitions of s_j and d_j, choosing y_j to be the inflow of the path P_j in the original preflow x yields a solution $\overline{y}$ of the system that fulfills all inequalities with equality. In the following, let $S := \{(y_1, y_3, \ldots, y_{\kappa-1}) : \overline{g}_j(y_j) + \overline{g}_{j+1}(s_{j+2} - y_{j+2}) \geqslant d_{j+1}$ for $j = 1, 3, \ldots, \kappa - 1\}$ be the set of vectors satisfying the inequalities and $D :=$ $[L_1, U_1] \times [L_3, U_3] \times \ldots \times [L_{\kappa-1}, U_{\kappa-1}]$ the set of vectors of allowed inflows. Since $\overline{C}$ was a T-cycle, the solution $\overline{y}$ mentioned above lies in $S \cap D^\circ$ and we are done if we can show that there also exists a solution in $S \cap \partial D$.[1]

For the sake of a contradiction, assume that there is no solution in $S \cap \partial D$ and, for $j \in \{1, 3, \ldots, \kappa - 1\}$, let

$$C_j := \{(y_1, y_3, \ldots, y_{\kappa-1}) : \overline{g}_j(y_j) + \overline{g}_{j+1}(s_{j+2} - y_{j+2}) < d_{j+1}\}$$

denote the set of points violating inequality j. Since $S = \mathbb{R}^{\kappa/2} \setminus \bigcup_{j \in \{1,3,\ldots,\kappa-1\}} C_j$ and we assumed that there is no solution in $S \cap \partial D$, the boundary ∂D must be contained in $\bigcup_{j \in \{1,3,\ldots,\kappa-1\}} C_j$. Since the functions $\overline{g}_i$ are convex, the sets $D \cap C_j$ are convex as well and we can find a hyperplane $H_j := \{y = (y_1, y_3, \ldots, y_{\kappa-1}) : \omega_j \cdot y = b_j\}$ with $\omega_j \cdot \overline{y} < b_j$ that separates $D \cap C_j$ from $\overline{y}$ for each $j \in \{1, 3, \ldots, \kappa - 1\}$. Thus, the set $P := \{y = (y_1, y_3, \ldots, y_{\kappa-1}) : \omega_j \cdot y \leqslant b_j$ for $j = 1, 3, \ldots, \kappa - 1\}$ is a polyhedron with $\overline{y} \in P$. In fact, since $\overline{y} \in P \cap (S \cap D)$, the polyhedron P must be a polytope enclosing $\overline{y}$ (otherwise, it would be possible to pass $S \cap \partial D$ following an extreme ray of P). Moreover, since $\overline{y}$ lies in the topological interior of P, we have $P^\circ \neq \emptyset$, which shows that P is a full-dimensional polytope in $\mathbb{R}^{\kappa/2}$. According to Lemma 7.20, a polytope of dimension $\frac{\kappa}{2}$ must have at least $\frac{\kappa}{2} + 1$ facets, whereas P was defined by only $\frac{\kappa}{2}$ inequalities, which yields a contradiction. Thus, there exists a solution in $S \cap \partial D$.

Hence, in terms of the original problem, we have now shown that there exists a feasible generalized preflow on G with flow value at least $\text{val}(x)$ for which the flow on at least one edge $e \in C$ is contained in $\{0, u_e\}$ (the flow value can have increased in case that the sink t is one of the nodes $\overline{v}_l$). Note that we only changed the flow on edges on C. Hence, by a repeated application of the above arguments, we finally obtain that there are no T-cycles left, which proves the claim. $\qquad \square$

Proposition 7.21 builds the foundation of the following main result of this subsection:

Theorem 7.22:
A maximum convex generalized preflow can be computed in $\mathcal{O}(3^m \cdot m)$ time.

Proof: Since we can restrict our considerations to preflows corresponding to basis structures as shown in Proposition 7.21, the theorem follows if we can show that,

---

[1] Here, D° denotes the topological interior of D and $\partial D := \overline{D} \setminus D^\circ$ the boundary.

given such a basis structure, we can compute feasible flow values on the edges in T that yield a maximum convex generalized preflow (or decide that the partition does not allow a feasible preflow) in $\mathcal{O}(m)$ time. We show that we can proceed similarly as in the traditional network simplex algorithm in order to reconstruct the flow that corresponds to a partition of the edge set. In doing so, we may discard partitions that may lead to feasible preflows. Nevertheless, we never discard partitions that lead to a maximum convex generalized preflow.

Clearly, we can discard partitions in which some node $v \in V$ that is incident only to L-edges and U-edges has a negative excess. Moreover, by Proposition 7.21, we can discard partitions that contain T-cycles and restrict our considerations to partitions in which the edges in T form a spanning tree of G.[2] Since T is a spanning tree, it contains the sink t. We designate t as the root of the tree T and seek to move the excess from each leaf of the tree towards the root in the following.

Each leaf v of the tree T is incident to exactly one edge $e \in T$. Let δ_v denote the excess at v generated by the L- and U-edges incident to v. We try to specify the flow on e in a way such that the excess at v becomes zero and will get transported towards the root node: If e is heading from v to some node w and $\delta_v < 0$, we discard the current partition since it does not allow a feasible preflow (since x_e is required to be non-negative, a negative excess will remain at v). If $\delta_v \geqslant 0$, we set $x_e := \min\{u_e, \delta_v\}$ in order to move the excess at v towards the root. Similarly, if e is heading from some node w to v, we set $x_e := \min\{u_e, g_e^{-1}(-\delta_v)\}$ if $\delta_v \leqslant 0$ in order to satisfy the demand at v and discard the partition if $\delta_v > 0$. In any case, since we have specified the flow on e, we can delete the edge from the tree and continue with the next leaf and so on. Note that this procedure maintains a non-negative excess at each leaf while creating the maximum possible excess at the corresponding adjacent inner node of the tree. Eventually, the flow on each edge $e \in T$ is determined (if possible) while creating the maximum possible excess at the root of the tree.

For each of the 3^m possible partitions (L, T, U) of E, we are able to find a node that is incident to exactly one edge in T efficiently by maintaining values $b(v)$ that correspond to the number of incident T-edges of a node $v \in V$ for which the flow value has not yet been fixed and a queue Q of nodes v with $b(v) = 1$. Whenever the flow value x_e of an edge $e \in T$ with end nodes v and w is fixed, we are able to decrease $b(v)$ and $b(w)$ and update Q in constant time $\mathcal{O}(1)$. Consequently, the reconstruction needs $\mathcal{O}(m)$ time, which proves the claimed total running time of $\mathcal{O}(3^m \cdot m)$. $\qquad\square$

---

2   Note that one can check in $\mathcal{O}(m)$ time whether a given graph contains an undirected cycle by using a depth-first search in the corresponding undirected graph.

Note that the maximum convex generalized preflow that is obtained by the above algorithm cannot be transformed into a maximum convex generalized flow without further assumptions: Suppose that there is a flow generating cycle (C, v) for some preflow x that creates s_v units of flow at v resulting in an excess of $\delta_v \in (0, s_v)$ at v. According to Lemma 7.7, we can describe the outflow of C at v by a convex function $\overline{g}$ of the inflow x_C of C at v that fulfills $\overline{g}(x_C) - x_C = s_v$. In order to get rid of the positive excess at v, we need to solve the equation $\overline{g}(x_C) - x_C = s_v - \delta_v$ in x_C, which is uncomputable in general even for a strictly increasing continuous convex function $\overline{g}$ since we do not have oracle access for the function $\overline{g}(x_C) - x_C$.

However, if we suppose that there is an oracle $\mathcal{A}$ that solves the above kind of equations in $\mathcal{O}(T_A)$ time, it is possible to find a maximum convex generalized flow on general graphs in $\mathcal{O}(3^m \cdot nmT_A)$ time as follows: For some partition (L, T, U) of the edge set E that implies a feasible preflow, consider a node $v \in V \setminus \{s, t\}$ with positive excess. Starting at $z := v$, we recursively follow some edge $e = (w, z) \in E$ with positive flow x_e and set $z := w$. Eventually, we either reach the source s or some node v' considered before, i.e., we obtain a cycle. In the first case, we find a feasible subtraction of Type I that either removes the excess at v or removes all flow on some edge on the underlying s-v-path. Note that this case may occur up to $\mathcal{O}(m + n) = \mathcal{O}(m)$ times and causes an overhead of $\mathcal{O}(n)$. In the second case, there is a minimum inflow x_P into the path P between v' and v such that the excess at node v and the flow on each edge on the path between v' and v remains non-negative. By incorporating the oracle $\mathcal{A}$, we can then either find a flow on the cycle such that the inflow into the path equals x_P while the excess at v' remains constant or we can reduce the flow on the cycle completely. Again, this case can occur up to $\mathcal{O}(m + n)$ times and causes an overhead of $\mathcal{O}(n \cdot T_A)$. This yields a total time requirement of $\mathcal{O}(nm \cdot T_A)$ per partition for converting the maximum preflow corresponding to the partition into a flow with the same flow value.

7.5.2 Acyclic Graphs

As shown above, according to Theorem 7.22, we can obtain a maximum convex generalized *preflow* on general graphs in $\mathcal{O}(3^m \cdot m)$ time. When restricting to acyclic graphs, we are able to turn preflows into flows efficiently, which can be incorporated into the proof of Theorem 7.22 in order to obtain a maximum convex generalized flow within the same time bound. This will be shown in the following corollary:

Corollary 7.23:
A maximum convex generalized flow in an acyclic graph can be computed in $\mathcal{O}(3^m \cdot m)$ time.

Proof: The algorithm used in the proof of Theorem 7.22 considers each possible partition of E into L, T, and U and computes a feasible preflow for this partition if possible. In acyclic graphs, the positive excess that might occur at any node $v \in V \setminus \{s, t\}$ stems from flows on s-v-paths according to Theorem 6.10. Thus, each of the computed preflows can afterwards be turned into a feasible flow by computing feasible subtractions of Type I on those paths that create the excess at the nodes $v \in V \setminus \{s, t\}$ with positive excess. This can be done in linear time $\mathcal{O}(m)$ as follows: Let $(v_1, \ldots, v_n)$ with $v_1 = s$ and $v_n = t$ denote a topological sorting of the nodes in V and let δ_v denote the excess of each node $v \in V$, which is non-negative for each $v \in V \setminus \{s, t\}$. Let i denote the maximum index with $i < n$ such that $\delta_{v_i} > 0$. Note that the positive excess at v_i stems from the ingoing edges $e \in \delta^-(v_i)$ only. Hence, as long as $\delta_{v_i} > 0$, we can reduce the inflow of some $e = (v_j, v_i) \in \delta^-(v_i)$ with $x_e > 0$, decrease δ_{v_i}, and increase δ_{v_j} appropriately until either $x_e = 0$ or $\delta_{v_i} = 0$. In the first case, we proceed with another edge in $\delta^-(v_i)$ with positive flow. In the second case, we again consider the maximum index $i < n$ with $\delta_{v_i} > 0$ until no such index exists. Eventually, we get rid of the positive excess at each $v \in V \setminus \{s, t\}$ by considering each edge at most once, which shows the claim. $\qquad\square$

7.5.3 Series-Parallel Graphs

We now restrict our considerations to the case of series-parallel graphs. Although Corollary 7.23 already provides an algorithm that solves CGMFP on series-parallel graphs exactly, it is possible to obtain a better running time by exploiting the inherent structure of the underlying graph. As it was shown in (Holzhauser et al., 2015b), the problem becomes solvable in $\mathcal{O}(2.83^m \cdot (m + n^2))$ time when using a more sophisticated approach of creating the basis structures. We present a revised approach that comes with an improved running-time of $\mathcal{O}(2.707^m \cdot (m + n^2))$ time and a simplified proof:

Theorem 7.24:
A maximum convex generalized flow in a series-parallel graph can be computed in $\mathcal{O}(2.707^m \cdot (m + n^2))$ time.

Proof: The idea of the algorithm is similar to the one presented in the proof of Theorem 7.22: For a given basis structure (L, T, U), we try to reconstruct the flow on the edges such that flow conservation is fulfilled at each node $v \in V$ if possible. Although the reconstruction procedure used in this proof has an increased running time of $\mathcal{O}(m + n^2)$ compared to the procedure used in the proof of Theorem 7.22, it yields a better overall running time since it can be interleaved with a traversal of a decomposition tree of the series-parallel graph in order to reduce the number of partitions that need to be considered within the algorithm.

Contraction Procedure:

We start by describing the new procedure for reconstructing the flow from a given partition of E into L, T, and U (or deciding that the partition does not correspond to a feasible flow). As a fourth kind of node, we introduce *unspecified* edges $e \in X$ for which the type will be determined in a later step of the algorithm. For each edge $e \in E$, we store two attributes: the interval in_e of potential inflows and the interval out_e of potential outflows. Obviously, for each $e \in L$, we get $in_e = out_e = [0,0]$ and, for each edge $e \in U$, we get $in_e = [u_e, u_e]$ and $out_e = [g_e(u_e), g_e(u_e)]$. Analogously, for $e \in T$ or $e \in X$, we have $in_e = [0, u_e]$ and $out_e = [0, g_e(u_e)]$ initially.

The algorithm is based on a fixed decomposition tree of the underlying series-parallel graph G. It repeatedly identifies *series trees* (maximal subtrees in which all inner nodes correspond to series compositions, cf. Figures 7.9a and 7.9c) or *parallel trees* (maximal subtrees in which all inner nodes correspond to parallel compositions, cf. Figure 7.9b) of the decomposition tree and contracts them into single edges. As an invariant that will be established in the creation process of the partitions below, we assume that every edge that corresponds to a leaf in a series tree is contained in X. Similarly, we assume that exactly one edge corresponding to a leaf in a parallel tree is in X while every other edge is contained in $L \cup U$.

Contraction of a series trees: Consider a series tree T with $k_T \geqslant 2$ leaves corresponding to edges $e_1, \ldots, e_{k_T}$ in the underlying series-parallel graph. As noted above, we can assume that the type of all of these edges is unspecified, so $in_{e_i} = [a_i, b_i]$ and $out_{e_i} = [c_i, d_i]$ for some values $a_i, b_i, c_i, d_i \geqslant 0$. In the original graph, the sequence $(e_1, \ldots, e_{k_T})$ corresponds to a path P of length k_T. Note that the flow on P is determined by the flow on one single edge on P. Similar as in the proof of Lemma 7.2, we can find a maximal interval of the form $[a, b]$ such that, for each $y \in [a, b]$, it holds that $x_{e_i} \in [a_i, b_i]$ for each $i \in \{2, \ldots, k_T\}$ in a flow x on P with $x_{e_1} = y$. We can, thus, replace the path P by a single unspecified edge $e \in X$ with $in_e = [a, b]$ and $out_e = [g_{e_{k_T}}(\ldots(g_{e_1}(a))\ldots), g_{e_{k_T}}(\ldots(g_{e_1}(b))\ldots)]$. In the decomposition tree, we similarly replace the series tree T by a new leaf corresponding to the edge e. Note that, if $in_e = \emptyset$, we can skip the given partition since it does not allow a feasible flow.

Contraction of parallel trees: Now consider a parallel tree T with $k_T \geqslant 2$ leaves corresponding to edges $e_1, \ldots, e_{k_T}$ in the underlying series-parallel graph. According to the above invariant, we can assume that there is exactly one index $j \in \{1, \ldots, k_T\}$ such that $e_j \in X$ and that there are two disjoint index sets I_L and I_U denoting empty and full edges, respectively, such that $I_L \cup I_U = \{1, \ldots, k_T\} \setminus \{j\}$. Note that the intervals in_{e_j} and out_{e_j} are of the form $in_{e_j} = [a_j, b_j]$ and $out_{e_j} = [c_j, d_j]$ while the intervals of every other edge e_i are of the form $in_{e_i} = [a_i, a_i]$ and $out_{e_i} = [c_i, c_i]$. Hence, the flow in the series-parallel subgraph G' that corresponds to the parallel tree T only depends on the flow on e_j, so we can replace T by a single leaf corresponding to an edge e with the intervals

$$in_e := \left[a_j + \sum_{i \in I_L \cup I_U} a_i, b_j + \sum_{i \in I_L \cup U_L} a_i \right]$$

and

$$out_e := \left[c_j + \sum_{i \in I_L \cup I_U} c_i, d_j + \sum_{i \in I_L \cup U_L} c_i \right].$$

Virtually, we replace the parallel edges $e_1, \ldots, e_{k_T}$ in G by the single edge e.

Before we derive the running time of the above contraction procedure, first note that each parallel tree and series tree can be determined by a single traversal of the decomposition tree in $\mathcal{O}(m)$ time in a preprocessing step. For each of the $\mathcal{O}(n)$ series trees, we need to compute the new intervals in_e and out_e, which results in an evaluation of $\mathcal{O}(k_T)$ outflow functions. Since each of these functions may result from the decomposition of other outflow functions from prior contraction steps, we obtain an overhead of $\mathcal{O}(n)$ per series tree. Similarly, each parallel tree T with k_T leaves causes an overhead of $\mathcal{O}(k_T)$. Since there are at most $\mathcal{O}(n)$ parallel trees (since the root of each parallel tree is either the root of the decomposition tree or a child of a series composition) and since each leaf in a parallel tree either corresponds to an edge in the original graph G or a series tree, the contraction steps of all parallel trees cause a total overhead of $\mathcal{O}(m + n)$. Hence, we obtain a total running-time of $\mathcal{O}(m + n^2)$ time for the reconstruction of the flow in each considered partition.

Partitioning procedure:

We now show how we can interleave the generation of the partitions into the contraction procedure described above while maintaining the claimed invariants. In particular, we are able to save some of the 3^m possible combinations by generating the partitions "just in time", i.e., prior to each contraction step.

Initially, we assume each edge to be unspecified. Suppose that the algorithm starts with the contraction of a series tree T with k_T leaves corresponding to the edges $e_1, \ldots, e_{k_T}$. As described above, we do not need to consider any partition of these edges but can replace them by a new unspecified edge. At some point in time (if the graph does not consist of a single path), we come across a parallel tree T with k_T leaves corresponding to the edges $e_1, \ldots, e_{k_T}$. At that time, all of these edges are unspecified. In order to fulfill the invariant required above, we need to consider different partitions of these edges into L, T, and U before the subsequent contraction step. According to Proposition 7.21, we can assume that *at most* one of these edges is of type T in a basis structure while the remaining edges are either empty or full. Equivalently, we can assess that *exactly* one edge remains *unspecified* while the remaining edges must be of type L or U. In total, we need to consider each of the k_T possible positions of the unspecified edge and, for every such position, each of the 2^{k_T-1} possible assignments of the remaining edges into L and U. We then contract the parallel tree into a single leaf, which is again unspecified as described above, and continue the procedure. An exemplary course of the overall procedure is depicted in Figure 7.9.

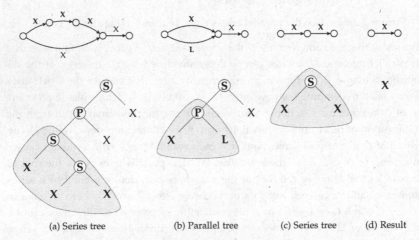

(a) Series tree   (b) Parallel tree   (c) Series tree   (d) Result

Figure 7.9: Iterative contraction of series and parallel trees in the algorithm. For each iteration, the upper figure shows the current graph and the lower one the current decomposition tree. The series tree in (a) can be contracted to a single edge without considering any further partitions, which yields the graph shown in (b). Afterwards, the two leaves of the parallel tree need to be assigned to L, U, or X such that the tree can be contracted. This yields the tree shown in (c), which can immediately be contracted into the single leaf shown in (d).

Complexity of the overall algorithm:

The bound on the time needed for the contraction steps has already been derived above. It remains to prove that the number of partitions that need to be considered can be bounded by 2.707^m. To this end, note that the number of partitions only increases when contracting a parallel tree.

Now consider a series tree T with $k := k_T$ leaves $e_1, \ldots, e_k$ prior to its contraction. Several contraction steps before, each e_i either corresponded to a parallel tree T_i with k_i leaves or to a tree T_i with only $k_i = 1$ leaf e_i that corresponds to an edge in the original graph (cf. Figure 7.10).

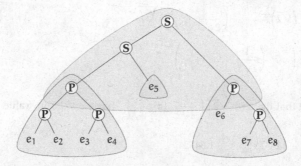

Figure 7.10: A series tree with $k = 3$ leaves and two parallel trees with $k_1 = 4$ and $k_3 = 3$ leaves. Before the algorithm contracts the series tree, it considers partitions for the two parallel trees and contracts these trees into single edges. It then contracts the series tree into a single edge.

Let $M(T)$ denote the number of partitions that need to be considered in order to process all of these trees T_i and the series tree T. As shown above, we get that

$$M(T) = \prod_{i=1}^{k} k_i \cdot 2^{k_i - 1}. \tag{7.2}$$

Since each of the considered trees is binary, it holds that the number n_i of nodes in each tree T_i is given by $2k_i - 1$ and that the number n_S of nodes in the series tree together with the nodes in the trees T_i is given by $n_S = \left(\sum_{i=1}^{k} n_i\right) + k - 1$. Note that, after the contraction steps of the trees T_i and the series tree T, the number of nodes in the decomposition tree is reduced by an absolute amount of $n_S - 1$. By substituting $k_i = \frac{n_i + 1}{2}$ in (7.2), we can bound the number $M(T)$ of partitions as follows:

$$M(T) = \prod_{i=1}^{k} \frac{n_i + 1}{2} \cdot 2^{\frac{n_i - 1}{2}} = \prod_{i=1}^{k} \frac{n_i + 1}{2\sqrt{2}} \cdot 2^{\frac{n_i}{2}} = 2^{\sum_{i=1}^{k} \frac{n_i}{2}} \cdot \prod_{i=1}^{k} \frac{n_i + 1}{2\sqrt{2}}.$$

By using the inequality of arithmetic and geometric means (cf. Cauchy (1821)), we get that

$$M(T) \leqslant \left(\sqrt{2}\right)^{\sum_{i=1}^{k} n_i} \cdot \left(\frac{\sum_{i=1}^{k}(n_i+1)}{2\sqrt{2}\cdot k}\right)^k$$

$$= \left(\sqrt{2}\right)^{\sum_{i=1}^{k} n_i} \cdot \left(\frac{1}{2\sqrt{2}}\cdot\left(\frac{\sum_{i=1}^{k} n_i}{k}+1\right)\right)^k.$$

For $z := \frac{\sum_{i=1}^{k} n_i}{k} \geqslant 1$, we further obtain that

$$M(T) = \left(\sqrt{2}\right)^{\sum_{i=1}^{k} n_i} \cdot \left(\frac{1}{2\sqrt{2}}\cdot(z+1)\right)^{\frac{\sum_{i=1}^{k} n_i}{z}}$$

$$= \left(\sqrt{2}\cdot\left(\frac{1}{2\sqrt{2}}\cdot(z+1)\right)^{\frac{1}{z}}\right)^{\sum_{i=1}^{k} n_i}.$$

It can be seen that the term $\sqrt{2}\cdot\left(\frac{1}{2\sqrt{2}}\cdot(z+1)\right)^{\frac{1}{z}}$ has a maximum value of

$$-4eW\left(-\left(2\sqrt{2}e\right)^{-1}\right) \approx 1.64524$$

for $z \geqslant 1$, where e is Euler's number and W denotes the Lambert W function. Hence, since $k \geqslant 2$, we get that

$$M(T) \leqslant 1.64524^{\sum_{i=1}^{k} n_i} = 1.64524^{n_S-k+1} \leqslant 1.64524^{n_S-1}.$$

Thus, in total, we only need to evaluate 1.64524^{n_S-1} partitions in order to remove $n_S - 1$ nodes from the decomposition tree. After contracting the series tree, we can repeat this procedure until the decomposition tree only consists of a single edge or of a single parallel tree. In any case, since we are interested in a maximum convex generalized flow, we do not need to consider any further partitions since it is clearly optimal to assign the remaining edges to U. Hence, since there are $2m - 1$ nodes in the decomposition tree, we get the following bound on the total number M of partitions that need to be considered:

$$M \leqslant \prod_{\text{series tree}} 1.64524^{n_S-1} = 1.64524^{\sum_{\text{series tree}}(n_S-1)} \leqslant 1.64524^{2m-1}$$

$$\leqslant 1.64524^{2m} \leqslant 2.707^m,$$

which shows the claim.   □

7.5.4 Extension-Parallel Graphs

As it was shown in Theorem 7.24, we can reduce the number of partitions that need to be considered from 3^m to 2.707^m by using a more sophisticated generation procedure for the problem of series-parallel graphs. When applying this algorithm to the convex generalized maximum flow problem on extension-parallel graphs, the number of partitions that need to be considered can be further reduced to 2.404^m as it is shown in the following corollary:

Corollary 7.25:
A maximum convex generalized flow can be computed in $\mathcal{O}(2.404^m \cdot m)$ time on extension-parallel graphs.

Proof: Assume that we apply the algorithm that was described in the proof of Theorem 7.24 to an instance of CGMFP on an extension-parallel graph G. Again, in order to bound the number of partitions that need to be evaluated, consider a series tree T with k leaves corresponding to the edges $e_1, \ldots, e_k$. According to the structure of extension-parallel graphs, it holds that *at most one* edge e_j among these edges results from a prior contraction of a parallel tree T_j with k_j leaves into a single edge. Again, since all of the considered trees are binary, it holds that the number n_j of nodes in T_j is given by $n_j = 2k_j - 1$ and that the number of nodes n_S in the series tree T together with the nodes in T_j is given by $n_S = 2k - 1 + (n_j - 1)$. Since $k \geqslant 2$, we get that $n_j = n_S + 2 - 2k \leqslant n_S - 2$. The number $M(T)$ of partitions that need to be considered in order to process the parallel tree (together with the series tree) is then given by

$$M(T) \leqslant k_j \cdot 2^{k_j - 1} = \frac{n_j + 1}{2} \cdot 2^{\frac{n_j - 1}{2}} \leqslant \frac{n_S - 1}{2} \cdot 2^{\frac{n_S - 3}{2}} = \frac{n_S - 1}{4} \cdot \left(\sqrt{2}\right)^{n_S - 1}.$$

For $z := n_S - 1$, we then get that

$$M(T) \leqslant \frac{z}{4} \cdot \left(\sqrt{2}\right)^{n_S - 1} = \left(\left(\frac{z}{4}\right)^{\frac{1}{z}}\right)^{n_S - 1} \cdot \left(\sqrt{2}\right)^{n_S - 1} = \left(\sqrt{2} \cdot \left(\frac{z}{4}\right)^{\frac{1}{z}}\right)^{n_S - 1}.$$

The maximum of $\sqrt{2} \cdot \left(\frac{z}{4}\right)^{\frac{1}{z}}$ is given by $\sqrt{2}e^{\frac{1}{4e}} \approx 1.55045$. As in the proof of Theorem 7.24, we finally get that the number M of partitions that need to be considered is bounded by

$$M \leqslant \prod_{\text{series tree}} 1.55045^{n_S - 1} = 1.55045^{\sum_{\text{series tree}}(n_S - 1)} \leqslant 1.55045^{2m-1}$$
$$\leqslant 1.55045^{2m} \leqslant 2.404^m,$$

which shows the claim. $\qquad\square$

7.5.5 Restricted Extension-Parallel Graphs

We close the study of graph classes with a special case of extension-parallel graphs that is solvable in polynomial time. As it was shown in Theorem 7.18, the problem CGMFP is $\mathcal{NP}$-hard to solve even if the underlying graph is restricted to be extension-parallel, i.e., if it is series-parallel but series compositions are only allowed in case that one of the two graphs consists of a single edge. We now show that CGMFP can be solved in linear time if we require that the *right hand side graph* (i.e., the graph whose source is identified with the sink of the other graph) in every series composition consists of a single edge. In the following, we refer to extension-parallel graphs with this additional restriction as *restricted* extension-parallel graphs.

Theorem 7.26:
A maximum convex generalized flow in a restricted extension-parallel graph can be computed in $\mathcal{O}(m)$ time.

Proof: Let $G = (V, E)$ be a restricted extension-parallel graph. The idea of the algorithm is to "pump" as much flow as possible into the graph in order to obtain a maximum preflow and to subsequently turn this preflow into a flow. For any series-parallel subgraph G' of G that corresponds to a node in the decomposition tree of G, we let $F(G')$ denote the maximum value of a convex generalized flow in G'. Starting from the leaves of the decomposition tree of G, these values $F(G')$ can be computed recursively as follows:

- If G' is a leaf of the decomposition tree corresponding to a single edge $e \in E$, we set $F(G') := g_e(u_e)$.
- If G' is the parallel composition of G_1 and G_2, we set $F(G') := F(G_1) + F(G_2)$.
- If G' is the series composition of G_1 and G_2, the right hand side graph G_2 must be a single edge e and we set $F(G') := g_e(\min\{F(G_1), u_e\})$.

Since each of the above steps requires only constant time $\mathcal{O}(1)$ and the decomposition tree contains $\mathcal{O}(m)$ nodes, this shows that we can compute the flow value $F(G)$ of a maximum convex generalized flow in G in $\mathcal{O}(m)$ time.

In order to compute the flow x_e on the edges of G in a maximum convex generalized flow x, we let $\text{out}_x(G')$ denote the outflow of each graph G' in the decomposition of G under x. Starting from the root of the decomposition tree, where we set $\text{out}_x(G) := F(G)$, each of these values $\text{out}_x(G')$ can be computed recursively as follows:

- If G' is the parallel composition of G_1 and G_2, we split the value $\text{out}_x(G')$ arbitrarily such that $\text{out}_x(G') = \text{out}_x(G_1) + \text{out}_x(G_2)$ and $\text{out}_x(G_1) \leqslant F(G_1)$, $\text{out}_x(G_2) \leqslant F(G_2)$.

- If G' is the series composition of G_1 and G_2, the right hand side graph G_2 must be a single edge e and we set $out_x(G_2) := out_x(G')$ and $out_x(G_1) := g_e^{-1}(out_x(G'))$.

Note that we always have that $out_x(G') \leqslant F(G')$ during the above procedure, so the splitting of $out_x(G')$ in case of a parallel composition is always possible and the procedure computes all values $out_x(G')$ in $\mathcal{O}(m)$ time. Afterwards, the flow x_e on each edge e can be computed from the value $out_x(e)$ obtained for the corresponding leaf of the decomposition tree as $x_e := g_e^{-1}(out_x(e))$. □

7.6 Integral Flows

We finally consider *integral flows* (i.e., feasible flows with integral in- and outflows for all edges) and assume that the outflow functions map integers to integers. Note that the $\mathcal{NP}$-completeness results from Theorem 7.17 and Theorem 7.18 remain valid for the case of integral flows. However, we are now able to derive a pseudo-polynomial-time algorithm for the problem on series-parallel graphs. In the following, let $U := \max_{e \in E} u_e$ and $\overline{U} := \max_{e \in E} g_e(u_e)$ denote the maximum possible inflow and outflow of an edge, respectively, which can be assumed to be integral as well without loss of generality.

Theorem 7.27:
A maximum integral convex generalized flow in a series-parallel graph can be computed in $\mathcal{O}(m^5 \cdot U^2 \cdot \overline{U}^2)$ time.

Proof: Consider a decomposition tree of G. For each component G' of this decomposition tree and for each value $x \in \{0, \ldots, m \cdot U\}$ and $y \in \{0, \ldots, m \cdot \overline{U}\}$, we compute the boolean function $A_{G'}(x, y)$, which is true if and only if an inflow of value x can produce an outflow of value y in G'.

Consider a leaf G' of the decomposition tree that corresponds to some edge e of the original graph G. For each $x \in \{0, \ldots, m \cdot U\}$ and $y \in \{0, \ldots, m \cdot \overline{U}\}$, we set $A_{G'}(x, y) :=$ TRUE if and only if $x \leqslant u_e$ and $g_e(x) = y$. If G' is the series composition of the two series-parallel graphs G_1 and G_2, we are able to achieve an outflow y with an inflow of x in G' if and only if there is some value $x' \in \{0, \ldots, m \cdot U\}$ that is both an outflow of G_1 and an inflow of G_2, i.e.,

$$A_{G'}(x, y) = \bigvee_{x' \in \{0, \ldots, m \cdot U\}} A_{G_1}(x, x') \wedge A_{G_2}(x', y).$$

Similarly, if G' is the parallel composition of the two series-parallel graphs G_1 and G_2, an outflow of y can be achieved with an inflow of x if and only if some amount y_1 of the outflow can be created with an inflow of x_1 in G_1 and the remaining outflow $y - y_1$ can be created with the remaining inflow $x - x_1$ in G_2. Hence, we get

$$A_{G'}(x,y) = \bigvee_{x_1 \in \{0,\dots,x\}} \bigvee_{y_1 \in \{0,\dots,y\}} A_{G_1}(x_1, y_1) \wedge A_{G_2}(x - x_1, y - y_1).$$

Note that there are $\mathcal{O}(m)$ nodes in the decomposition tree of G and we need to evaluate $\mathcal{O}((m \cdot U) \cdot (m \cdot \overline{U}))$ entries for each node. Clearly, for a single edge, each entry can be computed in constant time $\mathcal{O}(1)$. For the case of a series composition, we need to iterate over all possible values of x_1, which yields a complexity of $\mathcal{O}(m \cdot U)$ per entry. Finally, the evaluation of an entry for a node G' of the decomposition tree that corresponds to a parallel composition takes $\mathcal{O}((m \cdot U) \cdot (m \cdot \overline{U}))$ time. This yields the claimed running time of $\mathcal{O}\left(m \cdot (m \cdot U)^2 \cdot (m \cdot \overline{U})^2\right) = \mathcal{O}\left(m^5 \cdot U^2 \cdot \overline{U}^2\right)$. $\qquad\square$

7.7 Conclusion

We studied an extension of the generalized maximum flow problem in which the outflow of an edge is a strictly increasing convex function of its inflow. It turned out that the problem of computing a maximum convex generalized flow is strongly $\mathcal{NP}$-hard to solve even on bipartite acyclic graphs and weakly $\mathcal{NP}$-hard on extension-parallel graphs. For both cases and the case of preflows on general graphs, we presented exponential-time exact algorithms. Moreover, we showed that a flow decomposition similar to the case of traditional generalized flows is still possible and showed that the problem can be solved in pseudo-polynomial-time on series-parallel graphs for the case of integral flows. An overview of the results of this chapter is given in Table 7.1.

The model introduced in this chapter raises several interesting questions for future research. Since CGMFP was only shown to be weakly $\mathcal{NP}$-hard to solve on series-parallel graphs, it remains an open question whether a pseudo-polynomial-time algorithm exists also for the general case in which the flow is not restricted to be integral or whether the problem is actually strongly $\mathcal{NP}$-hard in this case. Furthermore, although the problem was shown to be $\mathcal{NP}$-hard to approximate, it remains open if and how approximate oracles could be used in order to obtain "almost feasible" solutions. Finally, although the running-time of the presented algorithms could be improved for the case of more simple graph classes, it may be possible to obtain faster algorithms by making even more use of the structure of such graph classes.

General Graphs	Acyclic Graphs	SP Graphs	EP Graphs
Theorem 7.13: Decomposable into m elementary subtractions	$\longrightarrow$	$\longrightarrow$	$\longrightarrow$
$\longleftarrow$	Theorem 7.17: strongly NP-complete to solve	$\longleftarrow$	Theorem 7.18: weakly NP-complete to solve
$\longleftarrow$	$\longleftarrow$	$\longleftarrow$	Theorem 7.19: NP-hard to approximate
Theorem 7.22: maximum preflow in $\mathcal{O}(3^m \cdot m)$ time	Corollary 7.23: maximum flow in $\mathcal{O}(3^m \cdot m)$ time	Theorem 7.24: maximum flow in $\mathcal{O}(2.707^m \cdot (m + n^2))$ time	Corollary 7.25: maximum flow in $\mathcal{O}(2.404^m \cdot (m + n^2))$ time
			Theorem 7.26: maximum flow in $\mathcal{O}(m)$ time on restricted extension-parallel graphs
		Theorem 7.27: maximum integral flow in $\mathcal{O}(m^5 \cdot U^2 \cdot \overline{U}^2)$ time	$\longrightarrow$

Table 7.1: The summarized results for the convex generalized maximum flow problem in Chapter 7. Implied results are denoted with gray arrows.

8 | Conclusion

In this thesis, we investigated the complexity and approximability of generalized network improvement and packing problems. In detail, we studied three extensions of the traditional maximum flow and minimum cost flow problem and revealed a strong connection to a novel variant of the bounded knapsack problem. For all of these problems, we both presented exact algorithms and investigated their approximability under involvement of a diverse set of graph classes.

As it became evident, extensions to the formulation of the traditional maximum flow or minimum cost flow problem that seem to be minor at the first glance turn out to have a significant impact on the complexity and approximability of the corresponding problems. Established combinatorial algorithms for well-known network flow problems turn out to be highly specialized to the inherent structure of these problems and cannot be directly applied to more general variants. The integrality assumption as a fundamental property of minimum cost flows could not be applied to any of the considered problems. When it was enforced to hold, it even made the budget-constrained minimum cost flow problem and the maximum flow problem in generalized processing networks $\mathcal{NP}$-hard to solve and approximate. Moreover, while efficient strongly polynomial-time algorithms are known both for the minimum cost and the maximum flow problem, the maximum flow problem in generalized processing networks turned out to be at least as hard to solve as any packing LP, which makes a strongly polynomial-time algorithm unlikely to exist. Finally, although the traditional maximum generalized flow problem can be solved efficiently, the counterpart considered in this thesis becomes strongly $\mathcal{NP}$-hard to solve and approximate.

Nevertheless, using more sophisticated approaches, we were able to adapt several results that are valid for the most fundamental network flow problems. In Chapter 4, we were able to extend the network simplex algorithm for the minimum cost flow problem to the more general case with an additional budget constraint. In addition, we were able to reduce the problem $\mathrm{BCMCFP}_{\mathbb{R}}$ to a sequence of $\widetilde{\mathcal{O}}(\min\{\log M, nm\})$ traditional minimum cost flow computations, so our algorithms benefit from the significant amount of research that lead to more and more advanced algorithms for the minimum cost flow problem in the past decades. Similarly, although the maximum flow problem becomes much harder to solve in the case of generalized processing networks as shown in Chapter 6, we were able to adapt a well-known result for the minimum cost flow problem to the case of our problem on series-parallel graphs. For the discrete versions of the budget-constrained minimum cost flow problem considered in Chapter 5, we observed an interesting connection of the problem on extension-parallel graphs to a novel variant of the bounded knapsack problem. This connection made it possible to derive efficient approximation algorithms for the budget-constrained minimum cost flow problem on extension-parallel graphs although the problem turned out to be $\mathcal{NP}$-hard to approximate on series-parallel graphs. This observation made once more obvious that network flow problems — just as knapsack type problems — are packing problems in their core.

In addition to the above results, we were able to show that one of the most fundamental theorems for network flow problems — namely the flow decomposition theorem — remains its validity for each of the considered problems (although the notion of "basic components" each flow decomposes into needs to be adapted). Hence, all of the considered problems can be seen as packing problems in which flows on such basic components are packed subject to a set of capacity constraints. This observation inspired the development of the generalized fractional packing framework in Chapter 3 as an integration of Megiddo's (1979) parametric search technique into the fractional packing framework of Garg and Koenemann (2007). This generalized framework leads to fully polynomial-time approximation schemes for a large class of network flow problems for which the flow decomposition theorem translates into the containment of each flow in a polyhedral cone, whose dual cone can be separated efficiently.

As already mentioned in the corresponding chapters, all of the investigated problems raise several questions for future research. It seems worthwhile to put more effort into further research on exact and approximation algorithms for the two continuous problems $BCMCFP_{\mathbb{R}}$ and MFGPN. In particular, one might hope to be able to derive new solution methods for these two problems by adapting algorithms for the traditional minimum cost and maximum flow problem (as it was done with the network simplex algorithm for the problem $BCMCFP_{\mathbb{R}}$) and applying more advanced speed-up techniques such as parameter scaling or the usage of sophisticated data structures like dynamic trees. For both problem, it would be fruitful to compare the empirical performance of the presented combinatorial algorithms to their non-combinatorial counterpart. For the three $\mathcal{NP}$-complete problems $BCMCFP_{\mathbb{N}}$, $BCMCFP_{\mathbb{B}}$, and CGMFP, it may be reasonable to develop more advanced approaches in order to speed up the existing algorithms and to identify efficiently solvable and approximable special cases that are of importance in practice.

Bibliography

D. P. Ahlfeld, J. M. Mulvey, R. S. Dembo, and S. A. Zenios. Nonlinear programming on generalized networks. *ACM Transactions on Mathematical Software (TOMS)*, 13(4): 350–367, 1987.

R. K. Ahuja and J. B. Orlin. The scaling network simplex algorithm. *Operations Research*, 40(1-supplement-1):S5–S13, 1992.

R. K. Ahuja and J. B. Orlin. A capacity scaling algorithm for the constrained maximum flow problem. *Networks*, 25(2):89–98, 1995.

R. K. Ahuja, T. L. Magnanti, and J. B. Orlin. Network flows. Technical report, Alfred P. Sloan School of Management, 1988.

R. K. Ahuja, K. Mehlhorn, J. B. Orlin, and R. E. Tarjan. Faster algorithms for the shortest path problem. *Journal of the ACM (JACM)*, 37(2):213–223, 1990.

R. K. Ahuja, A. V. Goldberg, J. B. Orlin, and R. E. Tarjan. Finding minimum-cost flows by double scaling. *Mathematical programming*, 53(1-3):243–266, 1992.

R. K. Ahuja, T. L. Magnanti, and J. B. Orlin. *Network Flows*. Prentice Hall, 1993.

R. K. Ahuja, J. B. Orlin, P. Sharma, and P. T. Sokkalingam. A network simplex algorithm with o (n) consecutive degenerate pivots. *Operations Research Letters*, 30(3): 141–148, 2002.

W. W. Bein, P. Brucker, and A. Tamir. Minimum cost flow algorithms for series-parallel networks. *Discrete Applied Mathematics*, 10:117–124, 1985.

L. Blum. *Complexity and Real Computation*. Springer New York, 1998.

L. Blum, M. Shub, and S. Smale. On a theory of computation and complexity over the real numbers: $\mathcal{NP}$-completeness, recursive functions and universal machines. *Bulletin of the American Mathematical Society*, 21(1):1–46, 1989.

M. Blum, R. W. Floyd, V. Pratt, R. L. Rivest, and R. E. Tarjan. Linear time bounds for median computations. In *Proceedings of the fourth annual ACM symposium on Theory of computing*, pages 119–124. ACM, 1972.

H. Booth and R. E. Tarjan. Finding the minimum-cost maximum flow in a series-parallel network. *Journal of Algorithms*, 15:416–446, 1992.

C. Çalışkan. A double scaling algorithm for the constrained maximum flow problem. *Computers & Operations Research*, 35(4):1138–1150, 2008.

C. Çalışkan. On a capacity scaling algorithm for the constrained maximum flow problem. *Networks*, 53(3):229–230, 2009.

C. Çalışkan. A specialized network simplex algorithm for the constrained maximum flow problem. *European Journal of Operational Research*, 210(2):137–147, 2011.

C. Çalışkan. A faster polynomial algorithm for the constrained maximum flow problem. *Computers & Operations Research*, 39(11):2634–2641, 2012.

A. L. B. Cauchy. *Cours d'analyse de l'École royale polytechnique*. De l'imprimerie royale, Debure frères, 1821.

M. D. Chang, C. J. Chen, and M. Engquist. An improved primal simplex variant for pure processing networks. *ACM Transactions on Mathematical Software (TOMS)*, 15(1):64–78, 1989.

V. Chankong and Y. Y. Haimes. *Multiobjective Decision Making: Theory and Methodology*. Dover Books on Engineering Series. Dover Publications, Incorporated, 2008.

A. Charnes and W. W. Cooper. Programming with linear fractional functionals. *Naval Research logistics quarterly*, 9(3-4):181–186, 1962.

C. J. Chen and M. Engquist. A primal simplex approach to pure processing networks. *Management Science*, 32(12):1582–1598, 1986.

T. H. Cormen, C. E. Leiserson, R. L. Rivest, and C. Stein. *Introduction to algorithms*. MIT press, 3rd edition, 2009.

W. H. Cunningham. A network simplex method. *Mathematical Programming*, 11(1):105–116, 1976.

W. H. Cunningham. Theoretical properties of the network simplex method. *Mathematics of Operations Research*, 4(2):196–208, 1979.

G. B. Dantzig. Application of the simplex method to a transportation problem. *Activity analysis of production and allocation*, 13:359–373, 1951.

G. B. Dantzig. *Linear Programming and Extensions*. Landmarks in Physics and Mathematics. Princeton University Press, 1965.

I. Demgensky, H. Noltemeier, and H. C. Wirth. On the flow cost lowering problem. *European Journal of Operational Research*, 137(2):265–271, 2002.

I. Demgensky, H. Noltemeier, and H. C. Wirth. Optimizing cost flows by edge cost and capacity upgrade. *Journal of Discrete Algorithms*, 2(4):407–423, 2004.

E. W. Dijkstra. A note on two problems in connexion with graphs. *Numerische Mathematik*, 1(1):269–271, 1959.

E. A. Dinic. Algorithm for solution of a problem of maximum flow in networks with power estimation. *Soviet Mathematics Doklady*, 11:1277–1280, 1970.

P. A. Maya Duque, S. Coene, P. Goos K., Sörensen, and F. Spieksma. The accessibility arc upgrading problem. *European Journal of Operational Research*, 224(3):458–465, 2013.

J. Edmonds and R. M. Karp. Theoretical improvements in algorithmic efficiency for network flow problems. *Journal of the ACM (JACM)*, 19(2):248–264, 1972.

M. Ehrgott. *Multicriteria optimization*. Springer, 2nd edition, 2005.

S-C. Fang and L. Qi. Manufacturing network flows: a generalized network flow model for manufacturing process modelling. *Optimization Methods and Software*, 18:143–165, 2003.

L. K. Fleischer and K. D. Wayne. Fast and simple approximation schemes for generalized flow. *Mathematical Programming*, 91(2):215–238, 2002.

L. R. Ford and D. R. Fulkerson. Maximal flow through a network. *Canadian journal of Mathematics*, 8(3):399–404, 1956.

M. L. Fredman and R. E. Tarjan. Fibonacci heaps and their uses in improved network optimization algorithms. *Journal of the ACM (JACM)*, 34(3):596–615, 1987.

M. R. Garey and D. S. Johnson. *Computers and Intractability – A Guide to the Theory of NP-Completeness*. W. H. Freeman and Company, New York, 1979.

N. Garg and J. Koenemann. Faster and simpler algorithms for multicommodity flow and other fractional packing problems. *SIAM Journal on Computing*, 37(2):630–652, 2007.

G. V. Gens and E. V. Levner. Fast approximation algorithms for knapsack type problems. In *Optimization Techniques*, pages 185–194. Springer, 1980.

A. M. Geoffrion. Solving bicriterion mathematical programs. *Operations Research*, 15(1):39–54, 1967.

A. V. Goldberg and S. Rao. Beyond the flow decomposition barrier. *Journal of the ACM (JACM)*, 45(5):783–797, 1998.

A. V. Goldberg and R. E. Tarjan. A new approach to the maximum-flow problem. In *Proceedings of the eighteenth annual ACM symposium on Theory of computing*, pages 136–146. ACM, 1986.

A. V. Goldberg and R. E. Tarjan. Solving minimum-cost flow problems by successive approximation. In *Proceedings of the nineteenth annual ACM symposium on Theory of computing*, pages 7–18. ACM, 1987.

A. V. Goldberg, S. A. Plotkin, and É. Tardos. Combinatorial algorithms for the generalized circulation problem. *Mathematics of Operations Research*, 16:351–379, 1991.

M. Gondran and M. Minoux. *Graphs and Algorithms*. Wiley, New York, 1984.

M. Grötschel, L. Lovász, and A. Schrijver. *Geometric Algorithms and Combinatorial Optimization*, volume 2 of *Algorithms and Combinatorics*. Springer Berlin Heidelberg, 1993.

Y. Han, V. Pan, and J. Reif. Efficient parallel algorithms for computing all pair shortest paths in directed graphs. In *Proceedings of the fourth annual ACM symposium on Parallel algorithms and architectures*, pages 353–362. ACM, 1992.

D. S. Hochbaum and A. Segev. Analysis of a flow problem with fixed charges. *Networks*, 19(3):291–312, 1989.

M. Holzhauser, S. O. Krumke, and C. Thielen. On the complexity and approximability of budget-constrained minimum cost flows. *submitted to Information Processing Letters*, 2015a.

M. Holzhauser, S. O. Krumke, and C. Thielen. Convex generalized flows. *Discrete Applied Mathematics*, 190–191:86–99, 2015b.

M. Holzhauser, S. O. Krumke, and C. Thielen. Budget-constrained minimum cost flows. *Journal of Combinatorial Optimization*, 31(4):1720–1745, 2016a.

M. Holzhauser, S. O. Krumke, and C. Thielen. A network simplex method for the budget-constrained minimum cost flow problem. *submitted to European Journal of Operations Research*, 2016b.

M. Holzhauser, S. O. Krumke, and C. Thielen. Maximum flows in generalized processing networks. *Journal of Combinatorial Optimization*, pages 1–31, 2016c.

K. Huang. *Maximum Flow Problem in Assembly Manufacturing Networks*. North Carolina State University, 2011.

D. B. Johnson. A priority queue in which initialization and queue operations take $\mathcal{O}(\log\log d)$ time. *Mathematical Systems Theory*, 15(1):295–309, 1981.

R. M. Karp. A characterization of the minimum cycle mean in a digraph. *Discrete mathematics*, 23(3):309–311, 1978.

A. V. Karzanov. Determining the maximal flow in a network by the method of pre-flows. *Soviet Mathematics Doklady*, 15:37–45, 1974.

H. Kellerer, U. Pferschy, and D. Pisinger. *Knapsack Problems*. Springer, 2004.

V. King, S. Rao, and R. Tarjan. A faster deterministic maximum flow algorithm. *Journal of Algorithms*, 17(3):447–474, 1994.

J. Koene. Maximal flow through a processing network with the source as the only processing node. 1980.

J. Koene. *Minimal cost flow in processing networks: a primal approach*. PhD thesis, Mathematisch Centrum, 1982.

B. Korte and J. Vygen. *Combinatorial Optimization*. Springer, 2002.

S. O. Krumke and S. Schwarz. On budget-constrained flow improvement. *Information Processing Letters*, 66(6):291–297, 1998.

S. O. Krumke and C. Zeck. Generalized max flow in series-parallel graphs. *Discrete Optimization*, 10(2):155–162, 2013.

S. O. Krumke, M. V. Marathe, H. Noltemeier, R. Ravi, and S. S. Ravi. Approximation algorithms for certain network improvement problems. *Journal of Combinatorial Optimization*, 2(3):257–288, 1998.

S. O. Krumke, H. Noltemeier, S. Schwarz, H-C Wirth, and R. Ravi. Flow improvement and network flows with fixed costs. In *Operations Research Proceedings 1998*, pages 158–167. Springer, 1999.

E. L. Lawler. *Combinatorial optimization: networks and matroids*. Courier Corporation, 2001.

H-Y. Lu, E-Y. Yao, and L. Qi. Some further results on minimum distribution cost flow problems. *Journal of combinatorial optimization*, 11(4):351–371, 2006.

H-Y. Lu, E-Y. Yao, and B-W. Zhang. A note on a generalized network flow model for manufacturing process. *Acta Mathematicae Applicatae Sinica, English Series*, 25(1): 51–60, 2009.

N. Megiddo. Combinatorial optimization with rational objective functions. *Mathematics of Operations Research*, 4(4):414–424, 1979.

N. Megiddo. Applying parallel computation algorithms in the design of serial algorithms. *Journal of the ACM (JACM)*, 30(4):852–865, 1983.

S. Mittal and A. S. Schulz. A general framework for designing approximation schemes for combinatorial optimization problems with many objectives combined into one. *Operations Research*, 61(2):386–397, 2013.

R. Motwani and P. Raghavan. *Randomized Algorithms*. Cambridge University Press, 1995.

Y. Nesterov and A. Nemirovskii. *Interior Point Polynomial Algorithms in Convex Programming*. Studies in Applied Mathematics. Society for Industrial and Applied Mathematics, 1994.

J. B. Orlin. A faster strongly polynomial minimum cost flow algorithm. *Operations research*, 41(2):338–350, 1993.

J. B. Orlin. A polynomial time primal network simplex algorithm for minimum cost flows. *Mathematical Programming*, 78(2):109–129, 1997.

J. B. Orlin. Max flows in $O(nm)$ time, or better. Technical report, Massachusetts Institute of Technology, 2013.

C. H. Papadimitriou. *Computational complexity*. Addison-Wesley Publishing Company, Massachussetts, 1994.

C. H. Papadimitriou and M. Yannakakis. On the approximability of trade-offs and optimal access of web sources. In *Foundations of Computer Science, 2000. Proceedings. 41st Annual Symposium on*, pages 86–92. IEEE, 2000.

T. Radzik. Improving time bounds on maximum generalised flow computations by contracting the network. *Theoretical Computer Science*, 312(1):75–97, 2004.

W. Rudin. *Principles of mathematical analysis*, volume 3. McGraw-Hill New York, 1964.

H. M. Safer and J. B. Orlin. Fast approximation schemes for multi-criteria flow, knapsack, and scheduling problems. Technical report, DTIC Document, 1995.

A. Schaefer. *Netze mit Verteilungsfaktoren*. Hain, 1978.

A. Schrijver. *Theory of Linear and Integer Programming*. John Wiley & Sons, Chichester, 1998.

R. L. Sheu, M. J. Tin, and I. L. Wang. Maximum flow problem in the distribution network. *Journal of Industrial and Management Optimization*, 2(3):237, 2006.

M. Shigeno. Maximum network flows with concave gains. *Mathematical Programming*, 107(3):439–459, 2006.

F. R. Spellman. *Handbook of Water and Wastewater Treatment Plant Operations, Third Edition*. Taylor & Francis, 2013.

R. E. Tarjan. Dynamic trees as search trees via euler tours, applied to the network simplex algorithm. *Mathematical Programming*, 78(2):169–177, 1997.

K. Truemper. On max flows with gains and pure min-cost flows. *SIAM Journal on Applied Mathematics*, 32:450–456, 1977.

K. Truemper. Optimal flows in nonlinear gain networks. *Networks*, 8(1):17–36, 1978.

P. Tseng and D. P. Bertsekas. An ε-relaxation method for separable convex cost generalized network flow problems. *Mathematical Programming*, 88(1):85–104, 2000.

P. M. Vaidya. Speeding-up linear programming using fast matrix multiplication. In *Foundations of Computer Science, 1989., 30th Annual Symposium on*, pages 332–337. IEEE, 1989.

J. Valdes, R. E. Tarjan, and E. Lawler. The recognition of series parallel digraphs. *SIAM Journal on Computing*, 11:298–313, 1982.

L. G. Valiant. Parallelism in comparison problems. *SIAM Journal on Computing*, 4(3): 348–355, 1975.

L. A. Végh. Concave generalized flows with applications to market equilibria. In *Proceedings of the 53rd Annual IEEE Symposium on the Foundations of Computer Science (FOCS)*, pages 150–159, 2012.

L. A. Végh. Strongly polynomial algorithm for generalized flow maximization. *arXiv preprint arXiv:1307.6809v2*, 2013.

P. Venkateshan, K. Mathur, and R. H. Ballou. An efficient generalized network-simplex-based algorithm for manufacturing network flows. *Journal of Combinatorial Optimization*, 15(4):315–341, 2008.

I. L. Wang and S. J. Lin. A network simplex algorithm for solving the minimum distribution cost problem. 2009.

K. D. Wayne. *Generalized Maximum Flow Algorithms*. PhD thesis, Cornell University, 1999.

K. D. Wayne. A polynomial combinatorial algorithm for generalized minimum cost flow. *Mathematics of Operations Research*, 27(3):445–459, 2002.

L. A. Wolsey and G. L. Nemhauser. *Integer and combinatorial optimization*. John Wiley & Sons, 2014.

M. Ziegelmann. *Constrained shortest paths and related problems*. PhD thesis, Saarland University, 2001.

Glossary

Notation

$\mathbb{R}$	set of real numbers
$\mathbb{R}_{\geqslant 0}$	set of non-negative real numbers
$\mathbb{R}_{>0}$	set of positive real numbers
$\mathbb{Q}$	set of rational numbers
$\mathbb{Q}_{\geqslant 0}$	set of non-negative rational numbers
$\mathbb{Q}_{>0}$	set of positive rational numbers
$\mathbb{Z}$	set of integral numbers
$\mathbb{N}$	set of natural numbers (without zero)
$\mathbb{N}_{\geqslant 0}$	set of natural numbers (including zero)
$\emptyset$	empty set
$A \subset B$	A is a proper subset of B
$A \subseteq B$	A is a subset of B (so $A \subset B$ or $A = B$)
$A \cap B$	intersection of A and B
$A \cup B$	union of A and B
$|A|$	cardinality of A
$g(n) \in \mathcal{O}(f(n))$	g grows at most as fast as f
$g(n) \in \widetilde{\mathcal{O}}(f(n))$	g grows at most as fast as f (up to poly-logarithmic factors)
$g(n) \in o(f(n))$	g grows slower than f
$g(n) \in \Omega(f(n))$	g grows at least as fast as f
$g(n) \in \Theta(f(n))$	g grows exactly as fast as f
$\log_b$	logarithm to the base b
$\ln$	natural logarithm (base e)
sgn	sign function
A_{ij}	entry in i-th row and j-th column of matrix A
$A_{i\cdot}$	i-th row vector of matrix A
$A_{\cdot j}$	j-th column vector of matrix A

x_e	flow on edge $e \in E$
$\delta^+(v)$	set of outgoing edges of node $v \in V$
$\delta^-(v)$	set of ingoing edges of node $v \in V$
$E(v,w)$	set of edges between nodes $v, w \in V$
$\mathrm{val}(x)$	flow value of flow x
$\mathrm{excess}_x(v)$	excess at node $v \in V$ in flow x
c_e	cost of edge $e \in E$
u_e	upper capacity of edge $e \in E$
b_e	usage fee/upgrade cost of edge $e \in E$
γ_e	gain factor of edge $e \in E$
α_e	flow ratio of edge $e \in E$
$g_e(\cdot)$	outflow function of edge $e \in E$
C	largest absolute value of an edge cost (in statements about time complexities)
U	largest capacity of an edge (in statements about time complexities)
B	largest usage fee/upgrade cost of an edge (in statements about time complexities)
M	maximum of C, U, B, and m
$SP(m, n, C)$	running time of the fastest shortest path algorithm
$SP(m, n)$	running time of the fastest strongly polynomial-time shortest path algorithm
$MF(m, n, U)$	running time of the fastest maximum flow algorithm
$MF(m, n)$	running time of the fastest strongly polynomial-time maximum flow algorithm
$MCF(m, n, C, U)$	running time of the fastest minimum cost flow algorithm
$MCF(m, n)$	running time of the fastest strongly polynomial-time maximum flow algorithm

Acronyms

Index

Printed in the United States
By Bookmasters